Oil Panic and the
Global Crisis

Oil Panic and the Global Crisis

Predictions and Myths

Steven M. Gorelick

Stanford University

WILEY-BLACKWELL

A John Wiley & Sons, Ltd., Publication

Library of Congress Cataloging-in-Publication Data

Gorelick, Steven M.
  Oil panic and the global crisis : predictions and myths /
Steven M. Gorelick.
    p. cm.
  Includes bibliographical references and index.
  ISBN 978-1-4051-9548-5 (hardback : alk. paper)
  1. Petroleum industry and trade. 2. Petroleum reserves–Forecasting. 3. Energy consumption–Forecasting. I. Title.
  HD9560.5.G594 2010
  333.8′232–dc22

                                        2009028753

ISBN: 9781405195485

A catalogue record for this book is available from the British Library.

Set in 10.5 on 12.5 pt Times by SNP Best-set Typesetter Ltd., Hong Kong
Printed and bound in Malaysia by Vivar Printing Sdn Bhd

1 2010

Contents

Preface

Scientists and engineers have applied prediction methods that lead them to the conclusion that the depletion of global oil is nearing. The majority of audiences I have addressed on this topic support this notion. The technological community has been reluctant to abandon its assumptions and tools, and yet there are those with a very different perspective that presupposes plentiful oil whose production is limited merely by price and technology. These two communities do not see eye to eye.

There is merit to the positions of both communities, and neither approach is free of major assumptions. So I embarked on a project: to present the primary arguments and the data that support and refute the idea that the world is in the precarious position of running out of oil. My hope is to both inform and instill some doubt on the subject in those who are steadfastly convinced one way or the other. Finally, there may be some value in viewing the modern issue of oil depletion in the context of the history of exploitation of other natural resources. Perhaps the light of this history can help illuminate our energy path before there is panic in the dark.

Cautionary note: *Nothing in this book should be used as the basis for financial investment. Oil and other commodity prices fluctuate over time and are subject to short-term volatility as well as long-term trends in economic, political, and natural conditions.*

Acknowledgments

This book was made possible by the John Simon Guggenheim Memorial Foundation, which provided a fellowship that is most gratefully acknowledged. Supporting recommendations from Professors William Yeh (University of California, Los Angeles), Garrison Sposito (University of California, Berkeley), and Pamela Matson (Stanford University) were deeply appreciated. Research for this book and most of the writing was conducted while I was on sabbatical leave from Stanford University for three periods. I sincerely thank Professor Wolfgang Kinzelbach for hosting me at Eidgenössische Techniche Hochschule (ETH) Institute for Environmental Engineering in Zurich, Switzerland in 2005; Professors Andrew Barry and Marc Parlange for hosting me at Ecole Polytechnique Fédérale de Lausanne (EPFL) Ecological Engineering Laboratory in Lausanne, Switzerland in 2006 and for providing access to the Les Bois Chamblard retreat facility; and Professors Andrew Balmford and William Sutherland of the Conservation Science Group for hosting me at the University of Cambridge, England in 2007. Each of these individuals and their research groups provided stimulating intellectual environments. I also thank my terrific chief research assistant on this project, Lauren Tippets, and student assistants Jerastin Dubash and Kirstin Conti for their helping hands with documentation. I am extremely grateful to expert illustrator David R. Jones for his preparation of the many graphics in this book, and to Paul Stringer for his editorial fine-tuning.

The enthusiasm, astute comments, and skillful editing of Stephanie Knott have greatly enriched the quality of presentation of often complicated material. Thank you Stephanie. Public debates about oil depletion with my brilliant colleague Amos Nur were inspiring. Superb critical review comments were provided by Larry Goulder, Alyssa Gorelick, Khalid Aziz, Stephan Graham, Marco Einaudi, Sally Benson, Fikri Kuchuk, Ashok Belani, David Howell, Steven Pearlstein, Myron Horn, Anne Edelstein, and Neth Walker.

About Units

This is a book with a lot of quantitative information about oil and other natural resources. Those unfamiliar with the overall topic of Earth resources might find the following unit conversions useful.

Volume

1 US gallon	3.785	Liters (l)
1 Imperial gallon	1.2	US gallons
1 US gallon	0.8327	Imperial gallons
1 barrel (bbl) of oil	42	US gallons
1 barrel of oil	34.97	Imperial gallons
1 barrel of oil	159	Liters

Oil Production Rate

1 billion barrels per year	2.74	million barrels per day
1 million barrels per day	0.365	billion barrels per year

Automotive Fuel Economy

US fuel economy units miles per US gallon (mpg)	Metric fuel economy units liters per 100 kilometers (l/100 km)	
10	23.5	
15	15.7	
20	11.8	
27.5	**8.6**	US fuel economy standard created 1975
30	7.8	
35	**6.7**	US fuel economy standard created 2007
40	5.9	

US fuel economy units   Metric fuel economy units
miles per US gallon (mpg)   liters per 100 kilometers (l/100 km)

50	4.7
75	3.1
100	2.4

Notes: Miles per US gallon = 235.2 ÷ liters per 100 kilometers
           Liters per 100 kilometers = 235.2 ÷ miles per US gallon
           1 Imperial mpg = 1.2 US mpg

Mass and Weight

1 metric ton (tonne)	1,000	kilograms
1 metric ton (tonne)	2,204	pounds

Energy

1 barrel of oil equivalent (BOE)[*]	5,660	cubic feet of natural gas
1 BOE[*]	160	cubic meters of natural gas
1 billion barrels of oil equivalent (BBOE)[*]	5.66	trillion cubic feet of natural gas
1 tonne of coal[*]	3.8	barrels of oil
1,000 BTU	0.293	kilowatt hours
1,000 BTU	1,055	kilojoules

[*] approximate, depending on fuel type and composition

Getting Started:
What Do *You* Think?

Before reading this book, you might find it interesting to identify your initial thoughts about the nature of global oil depletion. To gauge your views, please circle whether you "agree" or "disagree" with the following statements:

1. The availability of global oil has diminished over the past 100 years.

    Agree   Disagree

2. Over the past 50 years, some globally traded, non-renewable Earth resources have become scarcer, as indicated by a decline in the rate of extraction and an increase in price.

    Agree   Disagree

3. The inflation-adjusted price of gasoline in the US (apart from taxes) will be higher 30 years from now.

    Agree   Disagree

4. Gasoline and diesel fuel will power most passenger cars 30 years from now.

    Agree   Disagree

5. Multination wars will be waged with the aim of controlling regional oil resources.

    Agree   Disagree

1

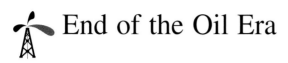 End of the Oil Era

Cause for Concern

The year 1859 was a double milestone in world history. Charles Darwin published his *Origin of Species*, and across the Atlantic, in a 33-state America, Edwin Drake sank the first US oil well in Titusville, Pennsylvania. Darwin offered the concept of the extinction of entire species as the backdrop for a potentially finite "Human Era," while Drake's discovery ushered in the "Oil Era," whose end, some speculate, is not very far off.[1,2] Since that year, oil has become the foundation of individual empires and a source of wealth for nations endowed with abundant reserves. Measured in antiquated units of 42-gallon barrels, oil is both a practical commodity and a tradable international currency. Oil is used to produce a wide diversity of products, such as fuel, plastics, paint, nylon, cosmetics, toothbrushes, and toothpaste. Our freedom of movement depends on oil for gasoline, a liquid that propels, pollutes, and has typically cost less than most bottled water. Oil's global abundance is ultimately unknown, yet the competition for control of this resource in the Middle East and fear about its future have been an impetus for war.[3]

That the world must run out of oil, perhaps some day soon, seems so obvious to most people that it is difficult to believe the topic is debated among scholars ranging from scientists to economists. After all, there is a finite amount of oil in Earth. That cannot be debated. Intuition tells us that scarcity is inevitable, given our societal history of consumption, our huge and continuing appetite for oil, and the fact that every developing nation relies on oil as a major stepping-stone to modernization. It stands to reason that

Oil Panic and the Global Crisis: Predictions and Myths. 1st edition. By Steven M. Gorelick.
Published 2010 by Blackwell Publishing, ISBN 978-1-4051-9548-5 (hb)

global energy consumption must be increasing with our ever-growing population, particularly in the emerging mega-industrial regions of China and India.

Common wisdom holds that since oil is a finite resource, its supply must be rapidly diminishing in the presence of clearly increasing demand, and the end of our days of oil must be on the horizon. Yet there have been predictions of the end of oil since it first became a common commodity. As early as 1916, the US Bureau of Mines stated, "… with no assured source of domestic supply in sight, the United States is confronted with a national crisis of the first magnitude."[4] A report commissioned by the US Department of Energy, called the Hirsch Report (2005), begins with the ominous warning, "The peaking of world oil production presents the U.S. and the world with an unprecedented risk management problem. As peaking is approached, liquid fuel prices and price volatility will increase dramatically, and, without timely mitigation, the economic, social, and political costs will be unprecedented."[5] A piece published in the journal *Science* in 2007 states, "The world's production of oil will peak, everyone agrees. Sometime in the coming decades, the amazing machinery of oil production that doubled world oil output every decade for a century will sputter. Output will stop rising, even as demand continues to grow. The question is when."[6] Similarly, a 2007 assessment by the US Government Accountability Office reported on the uncertainty of future global oil supply based on the premise that global oil production will peak and begin to decline "sometime between now and 2040," with the majority of cited studies suggesting that the production peak will likely occur by 2020.[7] Is this fear and doom-saying just another in a succession of false alarms?

Some experts say that there is plenty of oil. The essence of the idea that we will not run out any time soon was expressed in June of 2000 by former Saudi Oil Minister (1962–86) Sheikh Zaki Yamani. He claimed, "Thirty years from now there will be a huge amount of oil – and no buyers. Oil will be left in the ground. The Stone Age came to an end, not because we had a lack of stones, and the oil age will come to an end not because we have a lack of oil."[8] The Energy Information Administration (EIA), which is part of the US Department of Energy, says that only 4–7 percent of the world's original in-place liquid petroleum has been recovered.[9] Individuals ranging from oil company executives to energy consultants to academic economists firmly believe that any concerns about global depletion in the foreseeable future are premature for several reasons – oil is abundant, we have only consumed a fraction of the global oil endowment, technology to discover and extract new oil has consistently proven out, and the profit motive combined with the law of supply and demand will prevail.[10–13]

Why is our oil future so uncertain? What are the underlying data, analyses, and philosophies that lead to predictions of global oil depletion by some versus the conviction by others that the current state of alarm is unjustified and just crying wolf? Why is there any controversy at all? We can begin to answer these questions by considering the arguments supporting our intuition that the end of the Oil Era is near.

The Oil Era is a period of hundreds of years, contained in a longer period, during which global fossil fuel resources[14] are being consumed (Figure 1.1). Fossil fuels are the remains of ancient plants and animals. They represent a history of stored energy from the sun, which directly or indirectly gave them life and substance. Although fossil fuels took millions of years to form, humans are consuming them, and oil in particular, over a very brief span of Earth history. On the scale of thousands of years, looking back and to the future, "The consumption of energy from fossil fuels is thus seen to be but a 'pip,' rising sharply from zero to a maximum, and almost as sharply declining, and thus representing but a moment in the total of human history."[15]

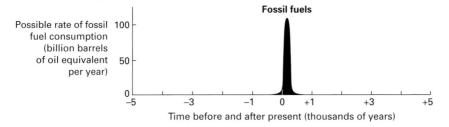

Figure 1.1   The consumption of fossil fuels considered over a 10,000-year horizon. The resource will likely be used during a relative instant of Earth history (after Hubbert, 1956 and 1981).[16]

Based on the presumption that the depletion of the world's oil resources is unavoidable, oil resource analysts have focused on four simple questions: How much oil exists to be exploited? What is the likely trend of new discoveries? What is the projected rate of global consumption? And, when will the end of the Oil Era arrive? Answering these questions is a subject of intense discussion in both the scientific literature and popular press. Many oil analysts have made estimates and projections. Cutting to the chase, most of these analysts have focused on the final question about when the end of oil will come, but they have framed the question in a slightly different way. The oil analysts focus not on when the last drop of oil will be pumped from the ground but rather the time when oil production will reach its peak ("peak oil"). It is their belief that the occurrence of peak oil production marks the beginning of the end, that is, the point when production can no longer keep up with demand. The argument goes that at the peak of oil production, the end is in sight, and it is urgent that a fundamental restructuring of our oil-based society begin.

So, when do analysts say that "peak oil" will occur? Surprisingly, the projections do not differ by much. The average collective estimate is that global peak oil production will occur before 2025, with the more pessimistic analysts suggesting that the peak has already occurred and we just do not know it, and the optimists pushing the date out to almost 2050. Remarkably, a great deal is made of the differences among the estimated times to peak oil production, and the debate among analysts is vigorous. But why is the exact year so important? The big message is that if they are correct, a key turning point in the nature of global industrial societies will occur within our likely lifetimes.

Hubbert's Curve

The general agreement among so many oil analysts regarding the time to peak oil production is not a tremendous surprise, because most use the same basic method for prediction. Although there are different flavors of the approach, they are based on the original method proposed by M. King Hubbert (1903–89), who initiated the modern-day scientific debate about oil depletion. Hubbert was a Texas-born geologist, oil company research scientist, and energy resource analyst. A respected scientist with a PhD in geology from the University of Chicago, he made durable contributions to the fields of both petroleum exploration and the study of natural subsurface water flow. After a 20-year career at Shell Oil and Shell Development companies, Hubbert joined the US Geological Survey (USGS) in 1963 and began a five-year teaching position at Stanford University. In 1973 he was appointed California Regent's professor at the University of California at Berkeley. Hubbert retired from academia in 1976, although he remained affiliated with the USGS.[17] He published more than 70 articles, and his work is still highly regarded and commonly referenced. Hubbert was famous during his lifetime, being elected to the National Academy of Sciences in 1955 and the American Academy of Arts and Sciences in 1957. He was the president of the Geological Society of America in 1962 and was awarded the Rockefeller Public Service Award in 1977. Hubbert was not only appreciated in scientific circles for his scholarly publications but also enjoyed attention in the press when it came to energy resources. After making predictions of the likely near-term depletion of US oil and natural gas as well as global oil resources, and suggesting that the development of nuclear energy was the best course of action, he testified before Congress on the bleak future of fossil-fuel energy resources.

The main theme in many of Hubbert's articles and monographs on energy resources was the fragility of our industrial global society that is so dependent on energy. He was fixated on the seemingly inevitable collision of finite Earth resources and the exploitation of those resources under the pressures of explosive global population growth. In a compelling 1949 article in the journal *Science*, Hubbert tied the consequences of exponential population growth to the general problem of fossil fuel depletion, considering oil, gas, and coal. Hubbert argued persuasively that even the habitable land required by society as we know it could not be sustained given a doubling of global population every hundred years, a 0.7 percent annual rate of increase that characterized population growth in the first half of the twentieth century. In his words,

> Such a rate is not "normal" as can be seen by backward extrapolation. If it had prevailed throughout human history, beginning with the mythical Adam and Eve, only 3,300 years would have been required to reach the present population. … In fact, at such a rate, only 1,600 years would be required to reach a population density of one person for each square meter of the land surface of the Earth.[18]

Throughout his career, Hubbert offered various forecasts of the decline in global oil supply, with published time-to-peak-oil predictions ranging from 1990 to 2000. His forecast peak dates were premature, but in the overall scheme of things, they do not differ dramatically from those made by modern-day energy analysts. Toward the end of his active career, Hubbert repeated the same somber message that he had put forth during prior decades,

> It is difficult for people living now who have become accustomed to the steady exponential growth in the consumption of energy from fossil fuels, to realize how transitory the fossil-fuel epoch will eventually prove to be when it is viewed over a longer span of human history.[19]

With his steadfast belief in exponentially increasing demand overtaking limited supply, Hubbert presented a quantitative method to represent the amount of any natural resource and its estimated rate of depletion. Hubbert's curve, as it is known, is a graph that shows the extraction of petroleum, or any non-renewable Earth resource, versus time. It is a bell-shaped curve, called a logistic curve,[20] similar in appearance to the bell-curve normal distribution commonly used in statistical analysis.

Hubbert used a straightforward formula that yields the curve as illustrated in Figure 1.2. The logistic-curve formula is a simple expression with three adjustable parameters (mathematical knobs) that control the slope, peak

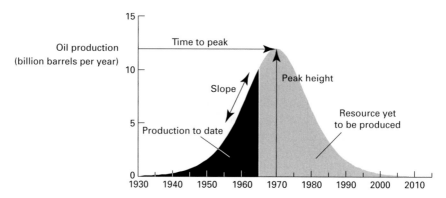

Figure 1.2   Generalized illustration of a logistic curve, showing its symmetric bell shape that Hubbert used to describe the rise, peak, and fall in production of a fossil fuel over time.

height, and time of the peak. The values of the parameters are adjusted to fit the historical production rates (data), which are matched by the curve since production began until production data are no longer available. With the constraint that the area under the curve represents the **resource endowment**, or total amount that can ultimately be produced plus the amount already produced, the formula is used to predict the future rate of resource production and depletion. The declining limb of the curve mirrors the rising limb. As Hubbert saw it, the use of any finite resource has a beginning, middle, and end. Indeed, it seems obvious that every finite commodity that is regularly consumed – from our life savings to our material supplies – must come to an end. Hubbert's curve reflects that commonly held belief.

Hubbert's approach was to take historical data of oil production over time and fit the logistic formula (his bell-shaped curve) to those data. The approach is attractive because anyone can reproduce it by fitting this or a similar bell curve to pre-peak production data. Hubbert observed that after oil was first extracted by wells in the 1860s, there was a rapid increase and then a marked decline in the discovery of new oil fields in the coterminous US, with the discovery peak occurring in the mid-1930s. He predicted that production would follow a similar decline. Figure 1.3 shows a logistic curve fit to historical oil production data through 2008 for US oil production in the lower 48 (coterminous) states, for which Hubbert estimated an oil endowment of 200 billion barrels.

The historical leg of the curve through 1956, when Hubbert made his original prediction, matches the oil production data. The early period of oil

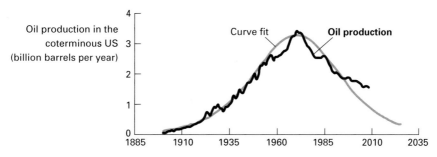

Figure 1.3   US lower 48 state oil production data and a curve fit using Hubbert's approach based on an estimated oil endowment of 200 billion barrels. (Data: EIA)

production and other resource utilization tends to display a rapid exponential increase. As time continued, oil production increased, but the rate of increase slowed until peak production (about 3.5 billion barrels per year for the lower 48 states) was reached. The date corresponding to this peak (actually occurring in 1970) is the time of maximum oil production, or "peak oil" production. Beyond peak oil, supplies presumably become depleted more rapidly than production from new discoveries can be brought on line. The curve after the peak falls back toward a level of nominal production. Eventually, the resource is exhausted when cumulative production nears the value of the oil endowment.

One might wonder how robust the curve-fitting procedure might be. That is, perhaps it is possible to fit the historical data with a variety of curves, each showing a different time to peak oil. It turns out that, even though the shapes of various curves that fit these data might be a bit different, the time of peak oil estimated using the approach does not vary by much. This is because there are two primary constraints controlling the curve-fitting process. The first constraint is that the historical data, typically representing the period before the peak, must be fit by the rising production limb of the curve. Only a limited subset of curves can match those historical data because they show a particular trajectory of increasing oil production. The second constraint is that the sum total of all production over time must equal the oil endowment. That total volume is a quantity that one must estimate independently of the curve-fitting procedure. It is this figure, the oil endowment, that is a subject of disagreement among oil analysts and one of the main sources of differing estimates of the time to peak oil. In essence, the trend of oil production largely dictates the uphill slope of the curve, while the total oil endowment controls the height and timing of the peak. As Hubbert himself stated in 1949,

Thus we may announce with certainty that the production curve of any given species of fossil fuel will rise, pass through one or several maxima, and then decline asymptotically to zero. Hence, while there is an infinity of different shapes that such a curve may have, they all have this in common: that the area under each must be equal to or less than the amount initially present.[21]

The time of peak oil based on fitting a logistic curve to the historical production data is not very sensitive to the independently estimated value used for the oil endowment. Figure 1.4 shows logistic curves fit to the US oil production data through 1956 but with three very different assumed oil endowments: 200, 300, and 450 billion barrels (only the first of these values was used by Hubbert). Although the peak value is very different in the three predictions, the time of the peak is not, in this case, 1971, 1981, and 1990. If Hubbert's estimate of the oil endowment is more than doubled, peak production is only delayed by 20 years. The various predictions based on different estimated oil endowment values all give similar times to peak oil. This is the main reason why oil analysts' predictions of the time of global peak oil typically vary by only about 20 to 30 years, even though they assume different oil endowment figures. This is also why the predictions do not differ significantly from Hubbert's predictions, first presented half a century ago.

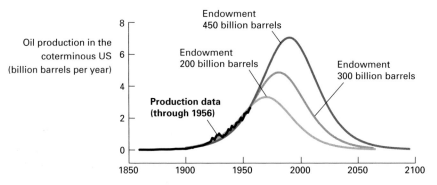

Figure 1.4   Curves based on Hubbert's approach that match US lower 48 state production data through 1956, when Hubbert first made his predictions, but assuming oil endowments of 200, 300, and 450 billion barrels: estimated peak oil production occurred in 1971, 1981, and 1990, respectively. (Data: EIA)

Between 1972 and 1976, Hubbert extended his analysis to global oil depletion. He made three estimates of the time of global oil depletion, with peak oil occurring in 1995, 1996, and 2000. Hubbert applied his approach using total global oil endowment figures ranging from 1.35 to 2.1 trillion barrels.

As displayed in Figure 1.5, a curve fit to the pre-1976 production data using the endowment value of 2.1 trillion barrels peaks at 35 billion barrels in the year 2000. Had Hubbert used the most recent worldwide oil endowment estimate by the US Geological Survey[22] of approximately 3 trillion barrels, the projected peak would occur in 2005. Using the two different global oil endowment estimates, the peak production values differ but the date of the peak is similar. With Hubbert's approach, the projected date of global peak oil production is rather immune to significant increases in the assumed oil endowment figure. From these analyses, it would appear that peak oil production is at hand.

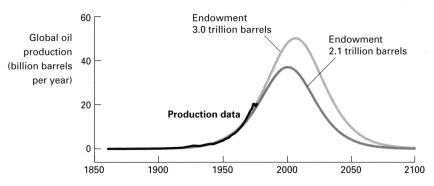

Figure 1.5   Global oil production data through 1976 when Hubbert made his final prediction of global oil depletion based on the highest endowment estimate he used of 2.1 trillion barrels. Also shown is a fit to production data using Hubbert's approach but assuming an oil endowment value of 3 trillion barrels. The time of peak oil does not change substantially under these scenarios. (Data: EIA and Hubbert (1969)[23])

Many analysts have followed in Hubbert's footsteps and made predictions of their own. There are various studies of the timing of peak oil production that show how different assumed oil endowment values and other factors, such as the growth of demand, affect the timing of global oil depletion. A 2004 study of global oil depletion in the journal *Energy* by Hallock and others looked at peak oil timing under a range of assumptions of oil endowment and future demand. Their model of oil depletion is more complicated than that of Hubbert, but the central idea is the same. They conclude that "global production of conventional oil will almost certainly begin an irreversible decline somewhere between 2004 and 2037."[24] Of equal concern, their study finds that demand will convert net-exporting countries into net consumers and that the number of exporting countries will fall from 35 in the present decade to between 12 and 28 by 2030.

The Appeal of Hubbert's Curve

For scientists and engineers, there is great appeal to Hubbert's method. For one thing, scientists like to bring order to data in a quantitative fashion. Given a set of data over time, many, if not most, scientists would study those data by seeing if a formula can be used to characterize them and perhaps explain trends, interpolate between values, and make projections. Often, curves are fit to data to allow a better understanding of the underlying process or processes that give rise to the data, such as trends suggesting exponential growth or decay. In engineering, physics, chemistry, and biology, formulas often describe governing forces and effects. Mathematics is the language used to quantitatively describe governing processes, even if a graphical presentation of a mathematical model, such as Hubbert's curve, is employed.

Engineers and scientists like to have a mathematical basis for making predictions beyond their current data, and predicting the trajectory of resource depletion is the main benefit of Hubbert's approach. The predicted rise, peak, and fall of Hubbert's curve follows naturally from the way he thought about the consequence of exponential growth in resource use. The logistic bell-shaped curve has a form that is familiar to scientists, engineers, and, indeed, much of the general public. Many are intuitively comfortable with such a curve being used to fit and "explain" a pattern of consumption.

Perhaps the main scientific appeal of Hubbert's approach is the fact that it represents a type of mass balance. One of the fundamental laws of science is conservation of mass, or in the case of oil measured in barrels, conservation of volume. There is only so much oil that can be extracted from Earth. Given that fixed volume of oil, how long it will last depends on its rate of extraction and consequent consumption. If the volume of new discoveries has already been incorporated into the oil endowment figure, then the oil in the ground is a "known," fixed volume that is simply waiting to be extracted. A fixed oil endowment subject to that oil being extracted over time is an expression of conservation of mass. Furthermore, the driving force behind oil consumption is assumed to be demand accompanying exponential growth of population and industry. This growth is the reason for the steep leading edge of the logistic curve. As long as global oil is plentiful, the effects of exponentially increasing production are not detrimental because there is ample supply to meet demand. However, as peak oil is approached, Hubbert and those using his method believe that demand will overtake the ability to extract oil. Both intuitively and mathematically, the future decline shown by Hubbert's curve appears to be the natural, inevitable result of the conflict between demand and a fixed, finite endowment. The result that there is a rapid increase in production followed by a symmetric, mirror-image decline is satisfying in the familiar sense that what goes up must come down.

Many scientists are comfortable with the Hubbert-curve approach because it describes a rate of depletion that is (1) consistent with conservation of mass, (2) based on a familiar mathematical and graphical form, (3) shows a trend with a leading limb that has a plausible behavioral underpinning tied to exponential growth in demand, and (4) is consistent with the expected declining production of a diminishing resource. Finally, scientists appreciate predictive models that are "robust," which means that uncertainty in the underlying data and parameters used in the predictive model do not greatly affect the results. As described above, even if the estimates of global oil endowment differ by a factor of two, the predicted oil production peak shifts by only two to three decades. This shift in timing is rather insignificant considering the anticipated consequences of global oil depletion.

Hubbert's Success

The successful demonstration of an approach goes a long way to convincing skeptics of its validity, particularly those in the scientific community. Hubbert did remarkably well with his early predictions. In 1956, Hubbert applied a precursor of his curve-fitting approach to oil production data in the coterminous US by fitting the historical production data by hand.[25] The data he used and his predictions are shown in Figure 1.6.

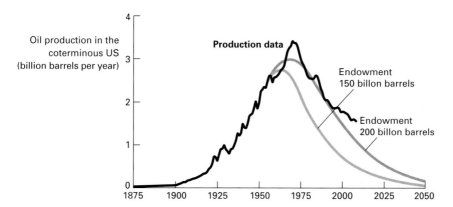

Figure 1.6 Hubbert's 1956 predictions of peak oil in the coterminous US modified to show production data through 2008 (after Hubbert (1956)[26] with production data from EIA).

Based on data through 1956, Hubbert attempted to predict the pattern of future oil production. His best estimate of the US oil endowment showed "peak oil" in the coterminous US occurring in 1965[27], and with his higher estimate of the endowment, the projected peak occurred in 1971. Compared with the actual long-term data collected after Hubbert made his predictions, which show a peak around 1970, Hubbert's estimates were surprisingly good. He successfully predicted the declining trend of oil production in the coterminous US, attributing that decline to the depletion of the available domestic resource. Given the right estimate of the oil endowment, the logistic curve can be used fairly well to describe both historical and post-peak production in the coterminous US. The success of Hubbert's approach in estimating the shape of the US oil production curve has been hailed as a remarkable achievement and has been used to support the applicability of his approach to global oil, coal, and natural gas resource analyses. Hubbert's prediction served as a wake-up call that oil is a finite resource and that its steadily increasing use would not be sustainable. Additional data gathered since Hubbert's time have provided the bases for numerous analyses supporting the conclusion that the end of the Oil Era is near.

US Oil Dependence Since Peak Production

Judging by the US experience, will a peak in global oil production matter? After the peak in US oil production in 1970, the US began to rely heavily on oil imports, and oil dependence has been a major focus of US foreign policy. The US maintains troops to defend the Middle East to insure against a global oil supply disruption. One cannot quantify the cost of potential loss of life that this service represents, but efforts have been made to determine the economic cost to the US government. Apart from the cost of war, there is a cost of having the military stand ready to defend the Middle Eastern supply.[28] Such annual costs are not reported by the US military; however, the US General Accounting Office[29] estimated the cost in 1990 of defending oil supply from the Middle East at $33 billion, which, adjusted for inflation, is $52 billion (in equivalent 2007 US dollars, 2007$). This cost corresponds well to a 2003 estimate made by the National Defense Council Foundation[30] of $49 billion, which, adjusted for inflation, is equivalent to about $55 billion (2007$).

Given that the US imports about 775 million barrels of oil per year from the Middle East and annually spends $50 billion per year to maintain a military force to defend that region, the standing armed force cost alone amounts to $65 per barrel. That "hidden cost" is equal to more than half of the average annual per-barrel oil price (adjusted for inflation – in 2007$) in any year through 2008. It is also more than double the $28 average annual price of oil since 1861.[31] The true cost of oil dependence encompasses far more than the price paid for gasoline. Should a peak in global oil production occur as it did in the US, oil-rich regions may well influence the global economy and the security concerns of all nations in ways that we have yet to experience.

Chapters Ahead

The concern about global oil depletion is not new, but the issue seems to draw attention primarily when oil and fuel prices climb. An appreciation of oil availability, supply, and demand issues in the US and the world is needed to evaluate opposing positions taken in the oil-depletion debate. Chapter 2 provides some key definitions and an overview of oil availability, production, and consumption. Alarm about the depletion of resources essential to society goes back at least 200 years. The resource-depletion debate and historical predictions of the exhaustion of natural resources, including oil, are discussed in Chapter 3. Given this historical context, additional arguments are presented that support the case for the world's running out of oil. Although there is compelling information that supports the case for global oil depletion, there are also counter-arguments that advocate the position that plenty of oil remains. These counter-arguments are made in Chapter 4. They focus on fallacies in the oil-depletion case and critique key assumptions underlying forecasts of global availability and demand. Finally, Chapter 5 applies some of the lessons learned from the examination of non-energy Earth resources to the analysis of global oil resources and explores important issues affecting our future reliance on oil.

Notes and References

1. Yergin, D. (1991). *The Prize: The Epic Quest for Oil, Money, and Power.* New York: Free Press, Simon and Schuster.
2. Darwin, C. (1859). *The Origin of Species*, reprinted in 1979 by Gramercy Books, Random House, New York: 461 pp.
3. Greene, D. L. (2009). "Measuring Energy Security: Can the United States Achieve Oil Independence?" *Energy Policy*. ISSN 0301-4215, DOI: 10.1016/j.enpol.2009.01.041 (in press).
4. Porter, E. D. (1995). "Are We Running Out of Oil?" American Petroleum Institute Policy Analysis and Strategic Planning Department, American Petroleum Institute, Discussion Paper #081.

5. Hirsch, R. L., R. Bezdek, and R. Wendling (2005). *Peaking of World Oil Production: Impacts, Mitigation, and Risk Management*, Department of Energy National Energy Technology Laboratory, February 2005.

6. Kerr, R. A. (2007). "The Looming Oil Crisis Could Arrive Uncomfortably Soon," *Science*, **316**, April 20, 2007: 351.

7. Government Accountability Office (2007). *Crude oil: Uncertainty about future oil supply makes it important to develop a strategy for addressing a peak and decline in oil production*, United States Government Accountability Office, GAO-07-283.

8. Fagan, M. (2000). "Sheikh Yamani predicts price crash as age of oil ends," *Daily Telegraph* (London, UK), June 25, 2000.

9. Sweetnam, G. (2008). "Long-term Global Oil Scenarios: Looking Beyond 2030," EIA 2008 Energy Conference, Washington, DC, April 7, 2008, Energy Information Administration.

10. Economides, M. J. (2007). "The Future of Peak Oil," *The Way Ahead* (produced by the Society of Petroleum Engineers), **3**(2).

11. Jackson, P. M. (2007). "Peak Oil Theory Could Distort Energy Policy and Debate," *Journal of Petroleum Technology* (produced by the Society of Petroleum Engineers), **59**(2).

12. Howden, D. (2007). "A world without oil," *The Independent* (London, UK), June 14, 2007: 1–2. "This scenario *of imminent global oil depletion* is flatly denied by BP, whose Chief Economist, Peter Davies, has dismissed the arguments of 'peak oil' theorists" (italics added).

13. Bahorich, M. (2006). "End of oil? No, it's a new day dawning," *Oil and Gas Journal*, **104**(31), August 21, 2006.

14. Fossil fuels are natural carbon-rich substances formed from the remains of plants and animals living millions of years ago. Found in Earth, these substances can be burned. They include coal, oil, natural gas, and peat.

15. Hubbert, M. K. (1949). "Energy from Fossil Fuels," *Science*, **109**(2823), February 4, 1949: 103–9 (p. 108).

16. Hubbert, M. K. (1956). "Nuclear energy and the fossil fuels," presented at the Spring Meeting of the Southern District Division of Production, American Petroleum Institute, San Antonio, TX, March 1956: Shell Development Company Publication No. 95; Hubbert, M. K. (1981). "The world's evolving energy system," *American Journal of Physics*, **49**(11): 1007–29.

17. "Tribute to M. King Hubbert," Letter to Members, National Academy of Sciences, 19(4), April 1990.

18. Hubbert, M. K. (1949). "Energy from Fossil Fuels," *Science*, **109**(2823) February 4, 1949: 105.

19. Hubbert, M. K. (1971). "The Energy Resources of the Earth," *Scientific American*, **225**, September 1971: 31–41 (p. 31).

20. Mathematically, the logistic curve is based on a simple, three-parameter formula in which, in our case, P is resource production, t is time, a is the peak height, b is the center or peak of the symmetric curve, and c controls the changing

slope, whose absolute magnitude is the same on both the rising and declining limbs. The area under the curve represents the total amount of the resource and is calculated as simply the product *4ac*.

$$\text{Production rate, } P = 4a\frac{\exp\left[-\left(\frac{t-b}{c}\right)\right]}{\left\{1+\exp\left[-\left(\frac{t-b}{c}\right)\right]\right\}^2}$$

21. Hubbert, M. K. (1949). "Energy from Fossil Fuels," *Science*, **109**(2823) February 4, 1949: 105.
22. US Geological Survey World Petroleum Assessment Team, (2000). *U.S. Geological Survey World Petroleum Assessment 2000—Description and Results.* US Geological Survey Digital Data Series DDS-60, 4 CD-ROMs; Ahlbrandt, T. S., R. R. Charpentier, T. R. Klett, J. W. Schmoker, C. J. Schenk, and G. F. Ulmishek (2005). "Global Resource Estimates from Total Petroleum Systems," AAPG Memoir 86, American Association of Petroleum Geologists, Tulsa, OK.
23. Hubbert, M. K. (1969). "Energy Resources," in *Resources and Man*, W.H. Freeman and Co. Chapter 8: 157–242.
24. Hallock, J. L. Jr., P. J. Tharakan, C. A. S. Hall, M. Jefferson, and W. Wu (2004). "Forecasting the limits to the availability and diversity of global conventional oil supply," *Energy*, **29**: 1673–96.
25. Hubbert, M. K. (1956). "Nuclear energy and the fossil fuels," presented at the Spring Meeting of the Southern District Division of Production, American Petroleum Institute, March 1956.
26. Ibid.
27. Ibid.
28. Davis, S. C. et al. (2008). *Transportation Energy Data Book*, Oak Ridge National Laboratory, ORNL-6981.
29. General Accounting Office (1991). *Southwest Asia: Cost of Protecting U.S. Interests*. Washington DC: US GAO/NSIAD-91-250, August 1991.
30. Copulas, M. R. (2003). *America's Achilles Heel – The Hidden Costs of Imported Oil*. Washington DC: National Defense Council Foundation, October 2003.
31. *BP Statistical Review of World Energy*, June 2008, http://www.bp.com/statisticalreview

2

The Global Oil Landscape

Introduction

Understanding global oil supply is complicated by opposing positions taken recently by major oil companies on the issue of "peak oil." On one hand, oil companies have advertised as fact that the world is rapidly depleting the remaining available oil. One TV advertisement by Chevron suggested that the remaining half-full global fuel tank may soon run dry, stating, "Some say that by 2020, we will have used half the world's oil. Some say we already have. Making the other half last longer will take innovation, conservation and collaboration." On the other hand, at the same time, there has been a declaration that peak production is not even on the horizon. Exxon Mobil ran a print op-ed piece entitled, "Peak Oil?" that argued, "Contrary to the theory, oil production shows no sign of a peak. ... The theory does not match reality, however. Oil is a finite resource, but because it is so incredibly large, a peak will not occur this year, next year or for decades to come."[1]

To obtain a clearer view of the global oil picture, some background information, primarily about oil's availability, is essential. This information includes the economics of oil supply and demand. Here we define some key terms and discuss the business of those supplying oil, the role of OPEC in dominating oil prices, estimates of the global quantity of oil and its regional consumption, gasoline and how its price relates to that of oil, and how price affects demand.

Oil Panic and the Global Crisis: Predictions and Myths. 1st edition. By Steven M. Gorelick.
Published 2010 by Blackwell Publishing, ISBN 978-1-4051-9548-5 (hb)

Definitions

Oil is derived from the fossil remains of prehistoric plants and animals. Despite media images to the contrary, oil is not formed from the decomposed remains of dinosaurs. Over the millions of years, or "geologic time," that oil was formed and trapped in sediments and rocks, dinosaurs either did not exist or were greatly outnumbered by very small aquatic plants and animals that are collectively termed "plankton." These mostly microscopic biota account for the greatest volume of organic matter that was available to form petroleum, from which oil is refined. It is estimated that each gallon of gasoline is the ultimate product of 90 metric tons (198,000 pounds) of ancient plant matter, which is 27,000 times the weight of the gasoline.[2]

Oil is a small fraction of **fossil fuels**, which are non-renewable (over the time frame of civilization), natural compounds consisting mainly of hydrogen and carbon (**hydrocarbons**) formed in the Earth's crust and including coal, oil, and natural gas. Oil is not the same thing as petroleum. **Petroleum** includes natural gas as well as crude oil and processed fuel products, whereas oil is the liquid that is refined to yield products such as gasoline, jet and diesel fuel, and lubricants.

There are some definitions related to oil availability that are central to the discussion of oil. A **resource** is "a concentration of naturally occurring solid, liquid, or gaseous material in or on the Earth's crust in such form and amount that economic extraction of a commodity from the concentration is currently or potentially feasible."[3] A more specific definition for petroleum resources is, "Concentrations in the earth's crust of naturally occurring liquid or gaseous hydrocarbons that can conceivably be discovered and recovered."[4] The **oil endowment** includes the estimated oil resource plus all of the oil that has been pumped from the ground. A subset of resources is reserves. A **reserve** is the portion of the resource that can "be economically extracted or produced at the time of determination."[5] A more targeted definition of petroleum reserves is, "The quantities of hydrocarbon resources anticipated to be recovered from known accumulations from a given date forward."[6] For the discussion in this chapter, the key distinction between the two quantities is that an oil resource is the presumed volume of all plausibly recoverable oil, while a reserve is the portion known to exist that can be profitably recovered with existing technology.

When oil is extracted from a reservoir by traditional methods (discussed later), only about one-third of the original oil in place is removed from the ground: two-thirds remain behind. Using modern methods of enhanced oil recovery, one-half to two-thirds or more of the original oil in place can be recovered. Cumulative production from an oil field will depend on its **recovery factor**, which is defined as the ultimate recovery divided by the initial oil in place. Greater oil recovery can be achieved at production costs considered high today but perhaps not so high in the future. Even if advanced technology is assumed to recover a greater fraction of the oil from a reservoir, it may be difficult to guess the future production costs. Therefore, the term **technically recoverable** oil is used for the portion of the oil resource "that is producible, using present or reasonably foreseeable technology, without any consideration of economic feasibility."[7]

Petroleum Composition and Energy Density

Petroleum is a complex mixture of organic compounds. At a given pressure and temperature, its components can be various kinds of liquids or gases, depending on its chemical composition. Crude oil is unrefined liquid petroleum. It is made up of many chains and rings of molecules consisting of hydrogen and carbon. The shorter-chain hydrocarbons are gaseous under normal surface conditions and are the most combustible, and even explosive, components of petroleum. The simplest substance is the lightest one, methane, which is the major component of natural gas and consists of one carbon atom surrounded by four hydrogen atoms. Methane is the primary component of so-called "dry gas" accumulations, those that occur in the absence of liquid oil. Methane is the cleanest-burning fossil fuel, since it has the lowest fraction of polluting carbon atoms. Methane is also associated with both oil accumulations and coal deposits. The commodity referred to as **natural gas** can occur alone in a subsurface reservoir or in mixed oil and natural gas systems. In mixed oil and natural gas accumulations, natural gas commonly consists mostly (70 to 90 percent) of methane but is in fact a mixture of free gas trapped above liquid oil and natural gas that is dissolved in crude oil. Natural gas accumulations can contain non-hydrocarbons, such as carbon dioxide, helium, nitrogen, and hydrogen sulfide. Other fuel gases, like ethane, propane, and butane, consist of two, three, or four carbon atoms, respectively, bonded in a chain and surrounded by hydrogen atoms. These methane-poor gases are the so-called "wet" gases and are frequently associated with naturally occurring oil, while "dry" gas is commonly generated from coal.[8]

Petroleum is refined to separate fuels from other products. Liquid hydrocarbon products, which contain many different compounds, consist of 5 to 19 carbon atoms. For example, gasoline is a mixture of hydrocarbon molecules with 5 to 7 carbon atoms, kerosene contains hydrocarbons with 8 to 13 carbon atoms, and diesel has hydrocarbons with 14 to 18 carbon atoms. Gasoline is 87 percent carbon by weight. Transportation fuels are lighter (less dense) than fuel oils (for example, home heating oil). Lubrication oils or greases consist of mixtures of hydrocarbons with anywhere from 18 to 40 carbon atoms in their chemical structures. Hydrocarbon compounds with 20 or more carbon atoms, such as grease, tar, and asphalt, form semi-solids at room temperature. All but the heaviest hydrocarbons are less dense than, and therefore will float on, water.

In terms of energy released from combustion, all refined fuels are not equal. For example, you would need five gallons of ethanol to give you the equivalent energy of three gallons of gasoline. The energy content per volume is a measure of **energy density** and can be expressed in BTUs (British thermal units) per gallon. Figure 2.1 compares the energy density of various fuels. The difference in energy density between diesel and gasoline is a consequence of these liquids' different ratios of hydrogen to carbon atoms. The higher ratio in diesel versus gasoline means that more energy is released per molecule of fuel. Per volume, diesel fuel yields 7 percent more energy than gasoline. Ethanol (drinking alcohol and a fuel made from corn or sugar) yields only 60 percent the energy of gasoline even though its hydrogen-to-carbon ratio is almost 50 percent greater than that of gasoline. This is a consequence of ethanol's oxygen-rich chemical structure, which releases proportionately less energy upon combustion.

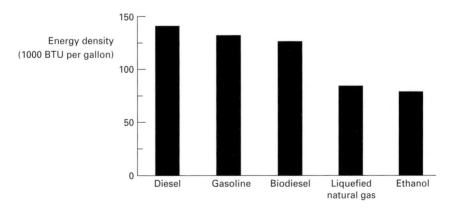

Figure 2.1 Comparison of the energy density of different fuels.[9]

Why a Barrel Is a bbl

Oil volume is measured in units of **barrels**, or **bbl**, an abbreviation that has an interesting history. Before the automobile was invented in 1885, there was no great use for gasoline. The value of oil was in the distilled kerosene that was burned for evening household lighting as an inexpensive replacement for other fuels that were clean enough for indoor use, such as whale oil. Other distilled oil components, such as gasoline, were discarded. When oil began to be produced in Pennsylvania in the 1860s, the immediate logistical problem was transport. There were no pipelines for oil, but wooden barrels were a convenient vessel for its temporary storage and shipment. Although barrels were commonplace some 150 years ago, they varied in size, and the lack of a standard for oil proved problematic in the budding oil market. At the time, standardized 40-gallon barrels were readily available, being produced for grain and salt in Virginia and used by Pennsylvania farmers to store and transport whiskey. However, it was not until around 1866 that oil buyers and sellers in northwestern Pennsylvania agreed to sell oil by the gallon and allow its shipment in 42-gallon barrels; the two gallons extra "in favor of the buyer" was apparently instituted to assure buyers that they would not be short-changed in their purchase of 40 gallons of liquid, which could, in principle, leak or evaporate.

After its adoption, the barrel, as a unit of measure, was accepted by the Petroleum Producers Association (1872) and by the US Geological Survey and the US Bureau of Mines (1882). Amazingly, the 42-gallon barrel is still used as the standard measure today and is represented by the symbol **bbl**. But why bbl for barrel? This abbreviation was apparently coined by the Standard Oil Company, whose late 1800s to 1911 monopoly of the oil industry is legendary.[10] Standard Oil provided "standard" oil barrels (bl) that were painted its trademark blue, hence the second "b" for "blue" barrel and the abbreviation bbl used today.[11]

The Oil Business

For over 100 years, big companies have sold oil. They reliably supply an essential product that keeps the world economy running, so why are they so often held in contempt? Some insight into the popular perception of multinational oil companies is gained by looking at the history of their profits (see Figure 2.2). Large oil companies' greatest profits have been secured during times of global crisis, war, natural disasters, and political turmoil, even though their profit margins have been meager relative to other industries during periods of serenity. From 1985 to 1999, major oil companies' return on investment (ROI) averaged 2 percent less than that of All Manufacturing Companies – a US Census Bureau benchmark. However, from 2000 through 2007, the annual ROI of the major US-based energy-producing companies tracked by the US Department of Energy under its Financial Reporting System (FRS) averaged 7 percent higher than the manufacturing benchmark.[12] This period was one of multiple stresses on the US and its oil-dependent economy, including disruption in imported supply from Venezuela, the war in Iraq, hurricane Katrina, and unexpectedly strong oil demand from China. The period preceding these stresses saw production cuts by the oil cartel OPEC. Rather than compete with their rival, US-based oil companies have shared without malice or collusion in the benefit of OPEC's production-based price controls.

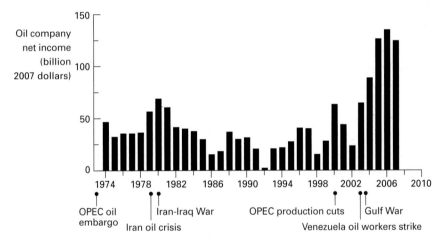

Figure 2.2 Combined net income of major energy companies reporting to the US Department of Energy and key corresponding events that drove up the price of oil and company net income above $40 billion per year.[13] (Data: EIA and DOE Financial Reporting System (2007 and 2008))

Oil is not big business, it is colossal business. Since the first US barrels of oil were tapped in 1859, the total global sale of oil of $40 trillion (2007$) through 2008 was more than ten times the entire US annual federal 2008 budget, or enough money to give over $5,000 to every man, woman, and child on Earth.[14] Of the top five 2008 Fortune 500 companies, oil companies accounted for 61 percent of revenues (Figure 2.3). From 2000 to 2008, Exxon Mobil ranked either first, second, or third in terms of Fortune 500 revenues, and from 2004 to 2007, its stock price more than doubled. The major

US-based energy-producing companies reporting to the US Department of Energy had total operating revenues in excess of $1.4 trillion, or 14 percent of the $10.6 trillion in revenues of all Fortune 500 companies in 2007.[15]

Oil companies have made big bets and have claimed big rewards. Oil companies risk huge sums to discover, develop, and produce oil and natural gas prospects. In 2007, the major US energy companies spent $14 billion on exploration, $54 billion on development, and $76 billion on production. Expensive offshore exploration has grown over the years and has increased the investment stake. For example, in 2006, Chevron, the second largest US oil company, completed a single exploratory well drilled beneath 7,000 feet of Gulf of Mexico waters to a total depth of 28,175 feet[16] at a cost of over $100 million. To develop the prospect, which lies about 175 miles off the Louisiana coast, an enormous investment is required – estimates range from three billion to tens of billions of dollars. On the other hand, the windfall from high oil prices through mid-2008 produced record profits. Net income of the 25 major US-based oil companies in 2007 was $125 billion or 29 percent of the Census Bureau's All Manufacturing Companies ($437 billion).[17] This was a record for the industry. From 2005 through 2008, Exxon Mobil, Chevron, and ConocoPhillips collectively accounted for most of the profits of the top five Fortune 500 companies.

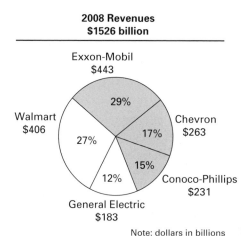

2008 Revenues
$1526 billion

Exxon-Mobil
$443

Walmart
$406

29%

Chevron
$263

17%

27%

15%

12%

Conoco-Phillips
$231

General Electric
$183

Note: dollars in billions

Figure 2.3   Revenues in billions of dollars of the top five Fortune 500 companies in 2008, showing the prominence of the oil industry. (Data: money.cnn.com, 2009)

The major US-based oil companies supplied 42 percent of US oil in 2007.[18] Private oil companies are massive, but in terms of oil production and reserves, they are dwarfed by state-owned (national) oil companies. In 2006, of the top ten oil companies in the world, 80 percent of production was by seven state-owned entities. As far as oil availability goes, just 3.8 percent of global reserves were held collectively by Exxon Mobil, BP, Chevron, ConocoPhillips, and Shell. But nine of the ten companies with the largest reserves were state-owned. Together these nine companies held 81 percent of the oil. All nine of those companies were controlled by OPEC member nations.[19,20]

OPEC

The most crucial component of the modern-day global oil production system is OPEC, the Organization of Petroleum Exporting Countries. OPEC was formed to maximize profits from the sale of oil by administering controls on world supply. Its members include Algeria, Iran, Iraq, Kuwait, Libya, Nigeria, Qatar, Saudi Arabia, the United Arab Emirates, Venezuela, and, as of 2007, Angola and Ecuador. Indonesia withdrew in 2008. Sitting on 70 percent of the world's 2008 oil reserves, this cartel has substantially controlled the price of oil. OPEC's 2008 oil reserves of 944 billion barrels were worth $56 trillion, assuming a price of $60 per barrel (below its 2007 average of $72 per barrel and well below its 2008 average of about $97 per barrel).[21] The value of OPEC oil exceeded that of all the goods and services produced in the world in 2008, the gross world product.[22] For each $11 per barrel rise in the price of oil, OPEC's reserves increase in value by $10 trillion.

Formed in 1960 from a core group, comprising Iran, Iraq, Kuwait, Saudi Arabia, and Venezuela, by 1971 the cartel had expanded to a group that controlled 52 percent of global production. In 2008, OPEC was responsible for 44 percent of global oil production. A relevant organization containing several OPEC members, known as OAPEC (the Organization of Arab Petroleum Exporting Countries), formed in 1968 and currently consists of OPEC members Algeria, Iraq, Kuwait, Libya, Qatar, Saudi Arabia, and the United Arab Emirates and non-OPEC members Egypt, Bahrain, and Syria.[23] Since its inception, OAPEC has been responsible for at least 58 percent of OPEC total production (Figure 2.4), and from 2000 to 2008 it accounted for two-thirds of OPEC production.

OPEC nations see maximum profits from the sale of oil compared to many other oil-producing nations because its members, particularly those in the Middle East, are generally the lowest-cost producers. For example, production costs have historically been in the range of $2 per barrel in the Middle East versus $17–25 per barrel in the US,[24] and oil production costs in the Middle East remain much lower than those in the US. Since its formation,

OPEC has taken in approximately $14 trillion (2007$), or more than one-third of all global oil revenue over all time. In 2008, OPEC net revenues from oil exports were $970 billion – almost ten times its revenues a decade earlier.[25]

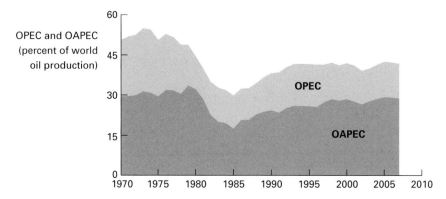

Figure 2.4   OPEC and OAPEC's share of global oil production since 1970.[26] (Data: EIA)

OPEC administers its control on the price of oil by placing quotas on the production of each of its members. In 2000, OPEC decided that if the price of the OPEC Basket of oil (discussed in more detail later) dropped below $22 per barrel for 10 consecutive days, then production would be curtailed. Even though OPEC also agreed to increase production if their oil Basket price rose above $28 per barrel, such action has been taken just once, in October of 2000. Meanwhile, after remaining around $25 per barrel through 2003, the average annual price of oil rose steadily to $54 per barrel in 2005 and climbed to about $97 per barrel in 2008 (with a temporary daily high value in excess of $145 per barrel).[27]

The relation between global demand and supply is the factor that dominates the price of oil. However, when it has had the opportunity, the OPEC cartel has introduced its non-competitive, production-based, price controls into the market. Economists have estimated that the competitive market price of oil would be under $20 per barrel if not for OPEC.[28] OPEC has administered control by limiting its production to a staggering degree. From 1973 to 1979, OPEC's average annual production was not very different from that during 2001 to 2008, even though today its member countries claim to hold over twice the global reserves that they held at the end of the prior period.[29] OPEC could continue producing oil for 80 years at its 2008 rate given its identified and profitable oil resources (without any discoveries). By comparison, the non-OPEC world can produce its reserves for 27 years. If just six OPEC countries – Saudi Arabia, Kuwait, United Arab Emirates, Venezuela,

Nigeria, and Iran – were to increase their annual production from 1 percent to 2 percent of their reserves, it would equal the combined production of the US, UK, Russia, Norway, Brazil, and China. Each set of countries accounts for 15 percent of global production, but this subset of OPEC nations holds six times the reserves of these non-OPEC nations.

OPEC has maintained the production capacity to satisfy global oil demand, but it is primarily interested in maintaining high oil prices by controlling production. As the price of oil in July 2008 fell from $145 per barrel toward $100 per barrel in September of 2008, OPEC decided to cut production by 520,000 barrels per day (less than 2 percent of total OPEC production). This intended cut was announced despite the fact that Saudi Arabia alone recently brought online the new Khusaniyan oil field with a production capacity of 500,000 barrels per day.[30] As oil prices fell to $35 per barrel in December 2008, OPEC announced additional production cuts bringing the total to 2.2 million barrels per day, expecting full implementation of the reduction by February 2009.[31] As the world economy declined into March 2009, OPEC's strategy of reducing production began to work – in one month the price of oil climbed by over 50 percent from its February 2009 low.

The global recession that began in 2008 created weaker demand for oil and most other commodities. Oil prices spiraled down beyond the immediate grasp of OPEC's production-based price controls. Their production cuts took time to implement. Even so, OPEC's stated intention was to limit its production and regain influence on the price of oil. Saudi Arabia considers $75 per barrel to be a "fair price."[32,33] The Secretary-General of OPEC, Abdalla Salem El-Badri, stated that OPEC "will not hesitate to take further measures to balance the market"[34] – in other words, cut production. He also claimed that low oil prices would lead to a failure of industry to invest in oil development and a consequent "supply crunch" by 2013.[35] OPEC's actions suggest that its goal is to generate higher and higher oil prices, not to stabilize them. However, it is not evident that OPEC member nations will abide by their individual production quotas in view of the needs of their individual economies.

OPEC argues that oil is the only significant commodity owned by many of its member nations and, as such, it must maintain what it considers a fair price. OPEC's attitude about its role in forcing the price of oil to near record levels in 2008 is one of denial of its manipulation of the market combined with an attitude that challenges the "consuming countries" to break free. In a 2007 public statement, OPEC noted:

> As it has done since it was established in 1960, the Organization will continue
> to work towards ensuring uninterrupted supplies to consumers at reasonable
> prices. But in the light of recent developments regarding the stated move away

from traditional fossil fuels – and oil in particular – OPEC Member Countries feel that they ought to review their future expansion plans. It would, in fact, make no sense for them to spend money unnecessarily on building or improving facilities when their customers are telling them they intend to minimize dependence on OPEC supplies.[36]

OPEC provides little reassurance to "consuming countries" with such statements. The global push toward alternative energy sources and conservation was encouraged and accelerated by the high price of oil in 2008. High prices served to heighten awareness and fear of the dependence on oil resources residing in what many people view as potentially vulnerable, if not precarious, parts of the world and coming from secretive suppliers who administer cartel economic power.

It should not be taken for granted that price stability is, in fact, best for OPEC to make money. Does price stability really maximize its profits? On the contrary, it is easy to imagine a world in which prices rise, plateau, and are then unexpectedly lowered by enhanced OPEC production. Consequently, non-OPEC investments in higher-cost and higher-risk oil exploration and production projects might stall and remain economically stranded as they suddenly become unprofitable. Because new technologies, including those that save energy, often are introduced at a premium price until they become widespread and products become commodity items, the strategy of a volatile oil market discourages the development of initially expensive alternatives to oil and conservation measures. Price uncertainty and instability are not necessarily detrimental to OPEC as long as it maintains control of production and generates long-term profits.

How Much Oil Is There? The USGS Assessment

Earth's resources, including oil, have been assessed by both governments and industry for many years. Figures for oil resources are compiled by the US Geological Survey and its sister agency, the Minerals Management Service. The US Geological Survey released its *World Petroleum Assessment 2000* ("Assessment"), a non-proprietary and freely available report.[37] In addition, oil production and consumption statistics are available from the US Department of Energy, Energy Information Administration (**EIA**), which regularly updates its online records.

The USGS Assessment is the best available and most comprehensive appraisal of the state of the world's ultimate oil resource, being the result of 100 person-years of effort by 41 individuals working from 1995 to 2000. The Assessment presents detailed evaluations of eight oil-bearing regions[38] throughout the world. The regions were subdivided into 128 geologic provinces with unique characteristics, and each province was evaluated for the amount of petroleum available. Not assessed were an additional 279 geological provinces where minor quantities of oil and gas have been found. The USGS notes that it is possible that significant petroleum exists in these and other areas not included in the Assessment.

The USGS 2000 Assessment team evaluated **conventional** sources of oil (liquid that is readily pumped from a well), natural gas, and natural gas liquids (**NGLs**), which are counted as a type of oil. NGLs are primarily the gaseous components recovered from natural gas reservoirs, and from natural gas that is naturally commingled with oil at depth in oil reservoirs but becomes liquid under surface conditions. The term NGL includes both natural gas plant liquids, which are processed separately as part of natural gas production, and "lease condensate," which is recovered from the oil. NGL obtained from gas processing plants is also referred to as liquefied petroleum gas (**LPG**).

The USGS Assessment[39] is quite valuable, but much of the report is very technical. For example, in their analysis of the Assessment results, the report uses terms like "future grown petroleum volume," "conventional petroleum endowment," and "discovery maturity."[40] Even professionals in many Earth science disciplines would be hard-pressed to know what these terms mean. However, descriptions of their analyses and assumptions are comprehensive, and USGS team members have made significant efforts to present their results at conferences and in publications.

The USGS global assessment includes both onshore regions and offshore regions; the offshore assessment incorporates areas up to 13,124 feet (4,000 meters) below the ocean surface. Not included are types of **unconventional** petroleum considered by the USGS to be too difficult and costly to exploit, such as oil sands, oil shales, gas hydrates, and heavy oil. These are forms of immobile or relatively immobile oil and natural gas that require special technology to exploit. Although not considered in the Assessment, in the past several years, unconventional oil sources have been making an entry as important energy resources. For example, oil sands, in which an immobile sand–oil mixture exists at shallow depths, are currently being exploited in Alberta, Canada. The increase in oil prices from 2003 through most of 2008, combined with new technology for oil-sand recovery, made production profitable. However, oil prices below those of 2008 have led to a slowdown in planned production.

In particular, the USGS Assessment looked at oil and natural gas resources that are expected to contribute significantly to proved petroleum reserves during the 30-year period 1996 through 2025. Remote environments, such as Antarctica, were excluded from the Assessment based on the assumption that exploration and recovery would not occur there in the Assessment's 30-year window. The USGS judged, with good reason, that future oil prices and technological developments are too uncertain beyond 2025 to extend its estimates of locating and producing oil beyond that date.

The USGS defined the term **total oil endowment** as consisting of four components: **cumulative production**, **remaining reserves**, **reserve growth**, and **undiscovered oil**. These terms are fairly self-explanatory. "Cumulative production" is the total reported volume that has been extracted since the 1860s. The term "remaining reserves" represents the known volume, yet to be extracted, that can be produced profitably. "Reserve growth," also known as field growth, is the amount of oil in addition to the known volume that will result from finding more oil in existing fields, improving recovery methods, and extracting oil that becomes economic to produce as prices or operating efficiency go up. "Undiscovered oil" is the volume likely to be encountered in promising environments (not including the 279 provinces that they did not study). Some data for the US were provided to the USGS by its sister agency, the Minerals Management Service. The USGS Assessment relied on commercial databases for its values of non-US world production and cumulative production (Petroconsultants, 1996, and NRG Associates, 1995).

The USGS estimate of total ultimately recoverable world oil, or oil endowment, was 3 trillion barrels (3.021 trillion barrels reported) – at least that is the expectation through 2025. The USGS also provided statistically based uncertainty bounds on their oil endowment estimate ranging from an optimistic 3.896 trillion barrels to a pessimistic 2.248 trillion barrels. Their best estimate is in agreement with an average of 14 industry estimates of 2.9 trillion barrels.[41] The 3-trillion-barrel global endowment is broken down by category in Figure 2.5. If natural gas liquids (NGLs) were included in

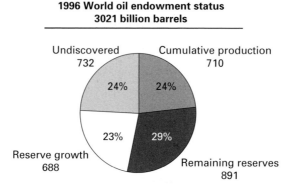

1996 World oil endowment status
3021 billion barrels

Undiscovered 732 — 24%

Cumulative production 710 — 24%

Reserve growth 688 — 23%

Remaining reserves 891 — 29%

Figure 2.5   USGS Assessment showing categories of the world's oil endowment (not including natural gas liquids) as of 1996 (including the US). (Data: USGS (2000)[42])

Figure 2.5, the global endowment would increase by 10.7 percent to 3.345 trillion barrels.

US oil endowment figures were estimated in a prior USGS study and were not officially part of their 2000 Assessment (but are included in Figure 2.5). The US share of the global oil endowment in 1996 was 12 percent, but because so much US oil has already been produced, only about half of the US endowment share remains untapped as reserves, reserve growth, and undiscovered oil. Figure 2.6 breaks down the US oil endowment into its components.

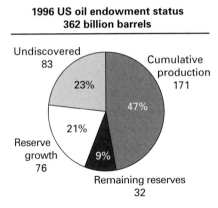

**1996 US oil endowment status
362 billion barrels**

Figure 2.6   Components of the US oil endowment estimate made by the USGS (1995).[43]

Roughly speaking, in 1996, the estimated world oil resources could be divided into four quarters. One quarter has been consumed (cumulative production), and known, profitably extractable reserves account for about another quarter. The remaining two quarters comprise "anticipated oil," with half coming from reserve growth and half from new discoveries.

From the USGS Assessment to 2009

Of course, since 1996, the amounts of reserves, production, and discoveries have changed. First, the USGS did not include in its assessment unconventional oil resources, yet in 2003, Canada's oil sands were recognized by *Oil and Gas Journal* (the industry standard) as bona fide reserves of 175 billion barrels. To keep the accounting straight, therefore, one must add 175 billion barrels to the USGS Assessment total, increasing the endowment to 3.2

trillion barrels (3.5 trillion barrels if NGLs are included). Second, reserves have been estimated to be 50 percent higher than in the global Assessment made for 1996. This difference in reserve estimates results from the USGS not including in their reserve estimates any oil from locations that they did not study in detail and omitting unconventional oil resources, as mentioned above. Thus, some of the oil that the USGS allocated to the "reserve growth" and "discovery" categories has been identified and deemed profitable such that it became 2009 "reserves." Third, cumulative production through 2008 was just over one trillion barrels. That is, about one-third rather than one-quarter of the estimated global oil endowment now has been consumed.

Lumping together reserve growth and new discoveries, the snapshot of world oil resources at the beginning of 2009 is shown in Figure 2.7. About one-third of the estimated oil endowment has been consumed, much less than one-half remains as reserves, and only one-quarter remains to be "found" through reserve growth or new discoveries.

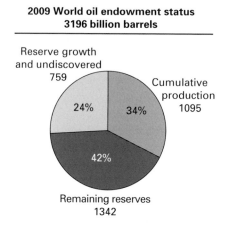

Figure 2.7   Global oil endowment components as of 2009, including Canadian oil sands. If 0.324 trillion barrels of natural gas liquids are added, the 2009 endowment value is about 3.5 trillion barrels. (Data: EIA and USGS)

If we assume that the oil endowment is a firm number around 3.2 trillion barrels, what does this mean in terms of global depletion? As an extremely rough and simplistic approach, one could estimate when the remaining oil would be consumed based on an assumed rate of consumption. Because about one-third of the oil endowment already has been consumed, about 2.2 trillion barrels remain. According to the EIA, world oil production was approximately 27 billion barrels in 2008. At that rate of production, which is not a readily justifiable measure of future production, the remaining 2.2 trillion barrels would be depleted in just over 80 years (2.2 trillion barrels divided by 27 billion barrels per year). Including NGLs in the calculation extends that time-frame by another 12 years, giving a total of 92 years. Under the assumed oil endowment and production values above, it took about 100 years to deplete the first third of the global oil endowment, and it might take less than 100 years to use the remaining two-thirds. If the average future worldwide production rate were double that of 2008, then depletion of a 2.2-trillion-barrel remaining endowment would occur in about 40 to 45 years.

Reserves

As discussed above, the oil endowment represents oil already consumed, oil in known accumulations, and oil likely to be found with existing technology through the year 2025. As a worst case for global oil depletion, we can consider only the oil that absolutely is known to exist in the world. Oil that is known to exist, has been clearly identified, and that can be extracted at a profit at the time of determination is called reserves. International oil reserve statistics are reported by the EIA as compiled by the *Oil and Gas Journal*. Figure 2.8 shows year-end 2008 oil reserves broken down by region of the world. Worldwide reserves at the end of 2008 were 1.34 trillion barrels (not including NGLs). At the 2008 rate of production, these reserves are sufficient for about 50 years.

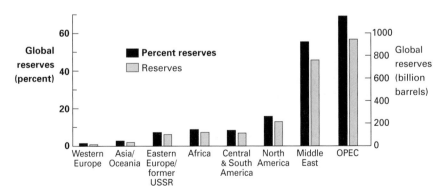

Figure 2.8 Regional percentage of global oil reserves at the end of 2008, showing regional breakdown, and global oil reserves (1.34 trillion barrels) breakdown. OPEC had 70 percent of global reserves in 2008. (Data: EIA and *Oil and Gas Journal*)

The regional breakdown used by the EIA differs from that in the USGS Assessment (Figure 2.9), but any way you look at it, the bulk of the world's oil is concentrated in the Middle East. As of 2008, the Middle East contained 56 percent of global oil reserves and was responsible for 45 percent of the global oil endowment. North America accounts for one-sixth of both reserves and the global endowment.

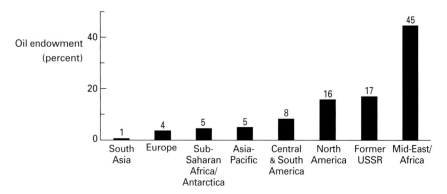

Figure 2.9   Breakdown of the global oil endowment based on the USGS Assessment including natural gas liquids. Here, North America does not include unconventional oil, oil sand. (Data: USGS (1995 and 2000))

Where Is Oil Produced?

Twenty-one countries produce over 85 percent of the world's oil (Figure 2.10). The greatest share of oil is produced by Saudi Arabia, followed by Russia and then the US. Global suppliers of conventional (liquid) oil are heavily weighted by countries in the Middle East, and all of the OPEC members are prominent in the list of significant oil producers.

Contrary to what one might expect, the countries with the largest oil reserves do not produce the most oil. Figure 2.11 shows the 20 countries that account for 95 percent of global oil reserves. Taken together, eight non-OPEC nations – the US, Mexico, Brazil, Russia, China, India, Norway, and the UK – have combined reserves below 11 percent of the world total, yet they produce 40 percent of the world's oil. Six OAPEC nations make up 41 percent of global reserves but supply just 25 percent the world's oil. The US ranks twelfth with less than 2 percent of global reserves but nevertheless is third in global oil production.[44]

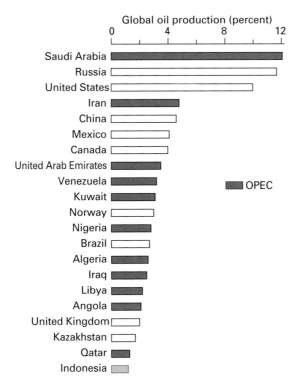

Figure 2.10   Oil production in 21 countries accounts for 85 percent of the world total (2007). Members of OPEC are shaded. Values include lease condensate, natural gas plant liquids, refinery gains, and other liquids. Indonesia is no longer in OPEC. (Data: EIA)

Where Is Oil Consumed?

Six countries consumed half of the world's oil in 2007: In descending order of consumption, the top six are: the US, China, Japan, India, Russia, and Germany (Figure 2.12). An additional 10 countries bring the share of total consumption to over 70 percent. The greatest user by far is the US, which consumes 24 percent of the global supply, more than twice that of any other nation, and yet the US represents only 4 percent of the world's population.

Viewed from a per capita standpoint, oil consumption looks very different, with Saudi Arabia, Canada, and the US being the largest consumers at 28, 25, and 25 barrels per person per year, respectively (Figure 2.13). There is a tremendous disparity in per capita oil consumption among different countries. In general, developed countries use more than about 10 barrels per person

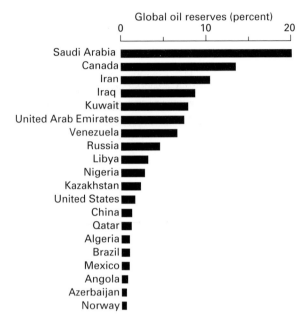

Figure 2.11 The oil reserves of 20 countries accounted for 95 percent of the world's total in 2008. (Data: EIA)

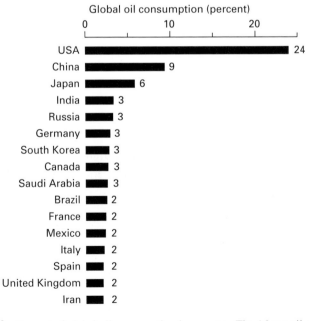

Figure 2.12 Percent of global oil consumption by country. The 16 top oil-consuming countries accounted for over 70 percent of all consumption in 2007. (Data: *BP Statistical Review of World Energy*[45])

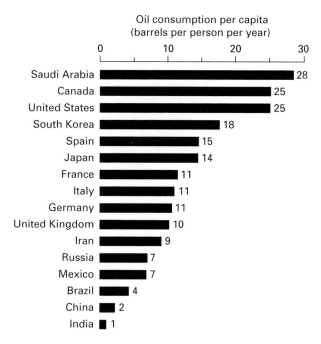

Figure 2.13 Oil consumption in barrels per person in 2007 rounded to the nearest barrel. (Data: consumption, *BP Statistical Review*; population, Economic Research Service, USDA)

per year, while underdeveloped nations, such as China and India, use less than two.

Oil Imports

Beginning in the early 1970s, the US went through a dramatic shift in dependence on imported oil. For the preceding 50 years, imports comprised about 10 percent of total US oil consumed. As shown in Figure 2.14, imports spiked during the 1970s oil crisis (OPEC oil embargo) and declined in the 1980s. Throughout the 1990s, imports climbed and since 2005 account for about two-thirds of all US oil consumed.

In 2007, the US imported oil from 46 different countries, although just seven provided 80 percent of imports. It may be surprising to some that two of the top five source nations of US imported oil are neither OPEC members nor in the Middle East: Canada is the US's largest, accounting for 19 percent of total US imports, and Mexico ranks third, providing 14 percent. The remaining three top nations responsible for imports to the US are indeed OPEC members: Saudi Arabia (15 percent), Venezuela (11 percent), and Nigeria (11 percent). Overall, OPEC accounted for about half of total US imports in 2007 (and 2008).

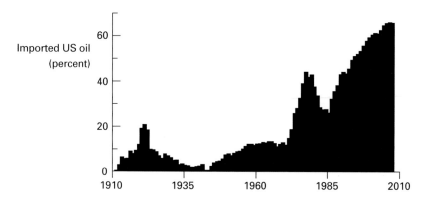

Figure 2.14   Percent of US oil that was imported through 2008. (Data: EIA)

The US dependence on OPEC for its imported oil has diminished over the years, peaking in the late 1970s at over 80 percent of imports and trending lower since then, stabilizing near 50 percent for the past decade (Figure 2.15). In terms of total US demand met by both domestic supply and imports, about 30 percent was satisfied by imports from OPEC in 2007 with just 6 percent from Persian Gulf nations. In addition to the US, the rest of the world, particularly Europe, is highly dependent on OPEC, as shown in Figure 2.16. Considered as a whole, the European OECD countries[46] imported over one-third of their oil from OPEC in 2007, with half of that coming from the Persian Gulf region. The great majority (85 percent) of Japan's oil demand is met by OPEC, almost entirely by Persian Gulf OPEC member nations. Although imports are reported for China, the data are uncertain. China imported about 42 percent of its oil in 2007 (and estimates range from 42 to 55 percent[47,48,49] for 2006). Historically, OPEC has supplied 30 to 45 percent of China's imports,[50] suggesting that 13 to 25 percent of its total oil demand is met by OPEC.[51,52]

Although much of the world's oil is supplied by OPEC, oil is a globally traded commodity, and imported oil can come from a variety of sources. Except for differences in quality, oil from one source is generally replaceable by oil from another source; that is, oil is a fungible commodity. As such,

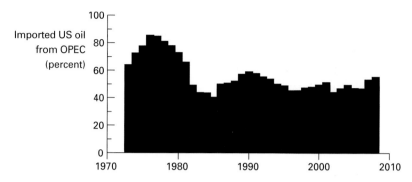

Figure 2.15   Percent of US oil imports obtained from OPEC through 2008. (Data: EIA)

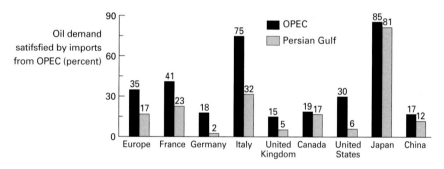

Figure 2.16   Oil demand satisfied by OPEC and Persian Gulf OPEC members in 2007. Europe is represented by OECD members. (Data: EIA; China value estimated. See text footnotes for data sources)

imported oil supply in the US, or any country, generally need not rely on exports from any particular OPEC or non-OPEC nation, as long as supply from an alternative source is available. Similarly, any oil-exporting nation can sell its oil as long as there is a buyer in the global marketplace. The price of the imported oil depends primarily on the relation between global supply and demand, with modest price variability related to regional factors affected by refining capability, oil quality, storage capacity, and markets.

After Oil Is Produced

Knowing what happens to oil once it is withdrawn from the ground and made into useful products is key to understanding global consumption. Different oils from different places in the world yield different product mixes. As an example, Figure 2.17 shows the volume of refined products from a typical barrel of US oil. Finished gasoline accounts for nearly one-half of each barrel. The total of all transportation fuels account for about 80 percent of oil, with petrochemicals, fuel gases, asphalt, lubricants, heating oil, and other products making up the remainder.

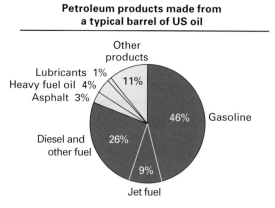

Figure 2.17   Petroleum products, mostly transportation fuels, made from a typical barrel of US oil.[53]

Surprisingly, if you sum the volume of petroleum products that are made from an average barrel of oil, the total comes to between 44.2 and 48.4 gallons (more than the 42 gallons you started with!), depending on how the oil is refined. This 5–15 percent bonus is called **processing gain**, which results from other petroleum products that are added to the oil during the refining process. In addition, in California, for example, another 5.7 percent of non-petroleum-based ethanol is added to make gasoline, bringing the products from the original 42-gallon barrel up to 49.6 gallons, which is an 18 percent gain in volume.[54]

Oil Production Versus Consumption

A comparison of global oil consumption and production over the past 25 years suggests, at first glance, that the world is consuming more oil than it produces (Figure 2.18). In fact, we consume what we produce, so it is important to understand the difference between reported crude oil production and consumption figures before we assume that there has been a continuing oil "deficit."

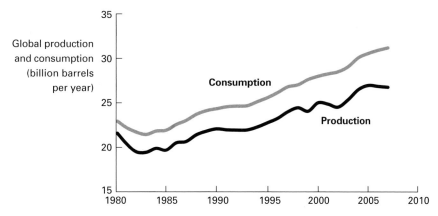

Figure 2.18 Global oil consumption exceeds production due to processing gains. Oil production values include "lease condensate" from natural gas that is naturally mixed in with the oil. Consumption includes processing gains, which include liquids added during the refining process and oil obtained from natural gas (natural gas liquid) that is produced and processed separately. (Data: EIA)

The key to this difference is processing gain. Worldwide, processing gains represent a significant "supply" of oil over the amount pumped from the ground and considered as production. In fact, since 2000, the world has consumed about 15 percent per year more oil than it has produced from oil wells (Figure 2.18). Crude oil production statistics reflect oil pumped from the ground, which consists of liquid oil plus "lease condensate." Reported oil production values typically include lease condensate. This requires a brief explanation. Some oil exists in the gaseous phase in deep natural oil reservoirs. When brought to the surface with the oil, this gas condenses into a liquid at atmospheric pressure. The condensed gas, now oil, is subsequently commingled with the crude oil that is simultaneously pumped from the ground. So crude oil production figures include the liquid oil plus the liquid from the condensation of the gas phase that is recovered.

On the other hand, oil consumption includes two components in addition to the volume of oil pumped from the ground: processing gains and **natural gas plant liquids**[55] (a portion of the NGLs). These liquids originate from natural gas production and not from oil production. The liquid is separated from the natural gas at downstream processing plants, and some is made into products like gasoline and jet fuel.

Worldwide, this seeming 15 percent excess of consumption over production is ostensibly enough to supply the current demand of 2.5 Africas, or one-third of all oil consumed annually in the US, or nearly half (44 percent) of European annual consumption. In fact, there is no excess. The important message is this: looking at production and consumption statistics, the fact that we consume more oil than we produce does not mean we are running an oil deficit. The discrepancy is accounted for when liquids derived from natural gas are considered.

Oil Quality

All crude oils are not the same: they can be light fluids versus near-solids at room temperature, vary in color from brown to black, and contain varying quantities of impurities. Because different crude oils contain different combinations of many hydrocarbons, the commercial products that each type of oil yields can be quite different. Pennsylvania-grade crude oil yields relatively small amounts of fuels like gasoline because it contains a significant amount of paraffin wax. But that grade of oil can be refined to make high-quality lubricants. On the other hand, oil from Prudhoe Bay, Alaska, contains less wax and yields more fuel.[56] Oil containing a higher proportion of long-chain hydrocarbon molecules yields relatively more lubricants and wax. Such oil typically contains a higher proportion of carbon relative to hydrogen atoms and is more likely to be a semi-solid at room temperature.

Crude oil also contains varying amounts of non-hydrocarbon impurities, such as sulfur, nitrogen, and oxygen. Such impurities not only affect the quality and price of the oil but also can be hazardous. For example, sulfur in the form of hydrogen sulfide, the gas with the familiar "rotten-egg" smell, is common in oil and natural gas and can kill a well operator within minutes.[57] As a major potential cause of acid rain (as sulfur oxides or SOx), the sulfur is removed during petroleum processing. Sulfur can make up 0.2 to 6 percent of the weight of crude oil. Oil that is rich in sulfur is less desirable and harder to refine. The oil industry labels oil with less than 1 percent sulfur "**sweet**," while oil with more than 1 percent sulfur is "**sour**."[58]

Oil Pricing by Quality

The price of crude oil is determined by both its quality and market factors. Over 160 different crude oils are traded globally. We can compare the quality and price of three common "benchmark" crude oils from Texas, the North Sea, and OPEC. West Texas Intermediate (WTI) is sweet light crude. It has only 0.24 percent sulfur and is refined to produce a large proportion of gasoline. Brent Blend, which reflects 15 oil fields in the North Sea, is also a sweet light crude, but with 0.37 percent sulfur is not as sweet (or light) as WTI. The OPEC Basket is an average of OPEC nation oils (except for Angola). The Basket accounts for the range from heavier sour oil originating in places like Dubai (United Arab Emirates) to better quality Algerian oil. It is high in sulfur, with 1.8 percent. The combined average quality has over four times the sulfur content of the Brent Blend and seven times the sulfur of WTI oil.[59] Figure 2.19 shows the prices of the WTI, Brent Blend, and OPEC Basket from 2000 to 2008. For that period, the average price premium for WTI over the OPEC Basket has been about 10 percent. Brent has traded at a premium of 5 to 10 percent over the OPEC Basket.[60]

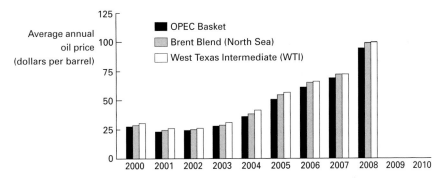

Figure 2.19 Annual average price (not adjusted for inflation, i.e., nominal dollars) of three types of oil of differing qualities. Note that although the spot (cash) market price of oil varied from about \$35 to \$145 per barrel during 2008, the average annual price remained below \$100 per barrel. (Data: EIA and OPEC)

In general, the prices of the oils of various qualities track together with the ups and downs of the global market. However, the differences in oil quality, ranging from sweet to sour and light to heavy, can account for a price differential of over 10 percent because refining costs and the refined product mix depend on the type of oil being processed.

Gasoline

What Determines the Price of Gasoline at the Pump?

There are four components of the cost of gasoline at the pump: (1) the price of the raw material, crude oil; (2) the cost (and profits) of refining the oil;

(3) the cost (and profits) of marketing and distributing the finished product; and (4) taxes. In the US, in May 2008 when the price of gasoline was $3.77 per gallon, three-quarters of the cost seen at the pump stemmed from the cost of the oil at $126 per barrel, about 5 percent from marketing and distribution, and the remainder (20 percent) from refining and taxes (Figure 2.20).[61] The average of federal and state taxes was about 38 cents per gallon in the US. Gasoline price changes are based primarily on the price of oil. For example, when oil was $33 per barrel in June 2004, 40 percent of the cost of gasoline was attributable to the price of crude oil. In November 2008, when the price of oil was $50 per barrel, 60 percent of gasoline's price was attributable to the price of oil.

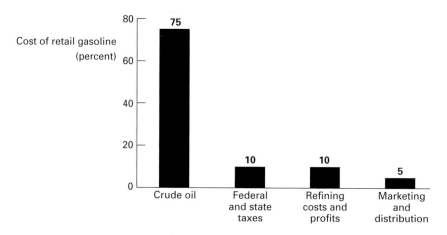

Figure 2.20   The cost components of gasoline in the US, when the total cost was $3.77 per gallon and oil sold for $126 per barrel in 2008. (Data: EIA)

As discussed above, the major cost component of gasoline is crude oil. Obtaining oil requires discovery, or finding costs, which involves exploration and development that add to proved reserves, plus the cost of getting the oil out of the ground, or lifting costs. These costs vary depending on the environment. Table 2.1 shows the finding and lifting cost breakdowns for 2006 and 2007 (in 2007$) for various regions of the world. The global average total cost was $26–$29 per barrel. The average cost in the Middle East for oil in 2007 was less than $14 per barrel. The costs of US offshore oil discovery and recovery were twice those of onshore costs in 2007. Oil from such offshore fields is economic only if prices are high or new technology reduces those costs.

Table 2.1   Finding and lifting (production) costs in 2007 dollars per barrel for crude oil in various regions of the world in 2006 and 2007[62]

	Finding cost 2006	Lifting cost 2006	Total 2006
Region			
United States			
Onshore	11.54	10.00	21.54
Offshore	65.49	7.52	73.01
Total US	15.95	9.37	25.32
Foreign			
Canada	19.89	8.90	28.79
Europe	23.41	8.57	31.98
Former Soviet Union	14.13	4.98	19.11
Africa	26.36	7.06	33.42
Middle East	5.41	14.92	20.33
Other Eastern Hemisphere	13.03	6.64	19.67
Other Western Hemisphere	43.87	5.73	49.60
Total Foreign	20.06	7.90	27.96
Worldwide average	**17.65**	**8.56**	**26.21**

	Finding cost 2007	Lifting cost 2007	Total 2007
Region			
United States			
Onshore	13.38	11.91	25.29
Offshore	49.54	8.92	58.46
Total US	17.01	11.25	28.26
Foreign			
Canada	12.20	10.41	22.61
Europe	31.58	10.35	41.93
Former Soviet Union	19.06	4.47	23.53
Africa	38.24	9.35	47.59
Middle East	4.77	8.61	13.38
Other Eastern Hemisphere	20.56	8.22	28.78
Other Western Hemisphere	30.30	5.92	36.22
Total Foreign	20.70	8.88	29.58
Worldwide average	**18.49**	**9.98**	**28.47**

The Price of Gasoline

Of common concern to the average consumer is not the price of oil, but the price of gasoline. How does the price of oil translate into the price of gasoline? There is a reasonable correspondence between the price of oil and the price of gasoline in the US, as shown in Figure 2.21, which presents inflation-adjusted prices of oil and gasoline since 1985. A more precise relationship between oil and US gasoline prices is revealed in Figure 2.22, which shows

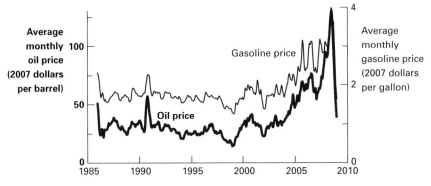

Figure 2.21   The correspondence between monthly average oil and gasoline prices in the US since December 1985. Prices adjusted for inflation.[63] (Data: EIA)

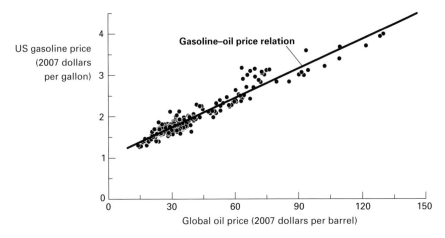

Figure 2.22   The correlation between oil and US gasoline prices for more than 20 years. For every $10 rise in the per-barrel price of oil, the price of retail gasoline increased by about 25 cents per gallon. Prices are adjusted for inflation. From 1919 to 2008, the average annual inflation-adjusted price (2007$) of gasoline has been $2.25 per gallon. (Data: EIA)

Table 2.2 Snapshot of March 2008 gasoline prices in different countries[64]

Country	Price per gallon
Norway	$8.73
United Kingdom	$8.38
France	$8.07
Germany	$7.86
United States	$3.45
China[65]	$2.61
Kuwait	$0.90
Saudi Arabia	$0.45
Iran	$0.40
Venezuela	$0.12

the linear relation of the two since 1985. For more than two decades, for every $10-per-barrel change in the price of oil, there has been about a 25 cent change in the price of retail gasoline.

Gasoline in the US is inexpensive compared with other oil importing nations. Price differences reflect taxes on or subsidies for gasoline use in different nations. The price of gasoline in much of Europe is about twice that in the US. Oil-rich OPEC nations provide enormous in-country subsidies, selling gasoline at a fraction of the price paid by consumers in the US. Amazingly, Venezuela sells gasoline for less than 5 percent of the price paid in the US. Comparative gasoline prices are shown in Table 2.2. In England, France, Germany, and other European countries, about two-thirds of the price of gasoline was tax in 2008, versus about 10 percent (averaging about 35 cents per gallon) in the US.

Gasoline Price Elasticity: What Happens When the Price Goes Up (or Down)?

Of great relevance to the discussion of global oil depletion is how the price of gasoline affects consumption. The demand for gasoline is affected by its price. Economists call this effect the price elasticity of demand. If the decrease in demand for a product is very responsive to a relatively small (say 10 percent) price increase, that product has high price elasticity of demand. On the other hand, high price inelasticity (low elasticity), or under-responsiveness, is a measure of the necessity of a product, or consumers' addiction to it, such that demand and consumption are only modestly reduced when the price goes up (or vice versa). Gasoline is a commodity that shows short-term highly inelastic behavior, meaning that an increase in the price is met by only a small decrease in consumption.

Historically, in the short term (perhaps a few years, before structural changes in consumption are made), it has been found that a 10 percent increase in the price of gasoline in the US results in a 2 percent decrease in consumption.[66] However, a 2008 study found that since 2000, short-term consumer response has changed dramatically.[67] For the period 2000 to 2006, the study found that a 10 percent increase in gasoline prices generated only about half a percent decrease (0.34 to 0.77 percent) in consumption, which is only about a quarter of the consumer demand decrease due to increased price that occurred from 1975 to 1980. During both of these periods, gasoline hit similar high prices (adjusted for inflation). The lower responsiveness to price in the recent period versus the 1970s may be due to more efficient vehicles, greater dependency on cars, and higher relative incomes such that consumers bought the same amount of gasoline even though the price had gone up. In any case, the low responsiveness is not surprising, since most individuals have fairly fixed transportation needs, and the cost of fuel must be absorbed as a necessary component of overall expenses. There are no convenient substitutes for gasoline, certainly in the short run, and US society is heavily dependent on cars. The results of this 2008 study of gasoline demand in response to price increases in the US are consistent with international oil short-term price elasticity of demand, which in 23 countries in North America, Europe, Scandinavia, and Asia translated to an average 0.5 percent demand decrease for a 10 percent price increase.[67]

On a longer-term basis, over many years, a 10 percent increase in the price of gasoline results in a 6 percent decrease in US consumption.[69] Long-term gasoline demand is still inelastic with regard to price, but much, much less so when compared to short-term consumer response. Over a greater length of time, consumers can adjust their commuting behavior (e.g., car pooling, more efficient cars) and reduce recreational consumption of gasoline (e.g., shorter trips, less use of cars). Consumers can, over time, restructure their buying habits in a way that they cannot do in the short term.

The value of the long-term price elasticity of demand for oil in the US corresponds closely to the price elasticity value for gasoline (i.e., for oil there is a 4.5 percent demand decrease for a 10 percent price increase). However, globally, the value with regard to oil is much lower than in the US. The UK, Denmark, Italy, South Korea, and Spain all show values of about 2 percent, with other countries like Australia, Norway, and Switzerland showing even lower values. These findings suggest that if current behavior is maintained, even over a period of years, much of the rest of the world is not likely to greatly reduce its consumption of oil as its price increases, at least for moderate price increases.[70]

Gasoline Price Variability

Oil companies made big profits with oil prices as high as they were from 2003 to mid-2008. These profits were justified by the enormous investments of oil companies in discovery and production. But are the rapid gasoline price variations at local gas stations, which sometimes occur daily, part of an overall profit-making scheme? No. There is a rather simple explanation for the rapid price variations seen at the gasoline pump, as noted by the California Energy Commission:

> Prices are set by what the station owner will have to pay for the NEXT delivery of gasoline. If prices are going up on the wholesale level, the station operator has to pay for what the next shipment will cost when it's delivered. If the delivery is going to cost more than what the dealer is charging at the pump now, they are going to lose money on the higher-priced new gasoline. So, the dealer has to increase his price whenever there is an increase in the wholesale price of the fuel to pay for that more expensive gasoline in the future.[71]

What about longer-term pricing of gasoline, the price of which seems to climb rapidly in response to oil-price increases and yet fall back slowly once oil prices decline? Here again there is frequent suspicion by the public that gasoline prices are being manipulated. Gasoline price manipulation has been discussed and studied by both those in government and academia. Figure 2.23

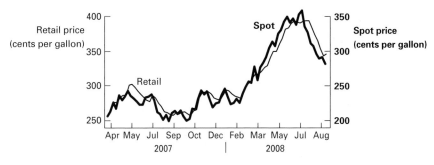

Figure 2.23 Retail US gasoline price and spot price, showing the correspondence between the two but the lag in the price decline of gasoline. (Data: EIA)

shows the market price for immediate delivery, or **spot price**, of US gasoline and the price of retail gasoline from 2007 to 2008. As prices increased, gasoline prices tended to track the increase,[72] but as prices fell, gasoline prices remained high, lagging the decline in spot prices. Economists call this lag effect "asymmetry of price pass-through," which differs from price gouging. Gouging has a technical definition: "… a situation where a seller attempts to extract a higher price (and profit) than would normally result from underlying supply and demand fundamentals."[73]

There is little regulation of the maximum price a station can charge for gasoline, although a few states have instituted "anti-gouging" laws. Federally, there is no law in place, but in 2007 the US House of Representatives passed an anti-gouging law that would make gouging a federal offense. That proposed law has been pending in the Senate since June 2007, as the White House then threatened a veto.[74] In response to concerns about price manipulation in the gasoline market, in 2005, the US Federal Trade Commission (FTC) issued a comprehensive analysis of gasoline price changes.[75] Their report cites simple supply and demand as the reason for the lag effect and the overall price behavior of gasoline.

In general, price gouging is uncommon and unsustainable for a commodity whose demand is relatively inelastic,[76] because if the price is too high, people buy their gasoline elsewhere, and the station does less business and hence makes less money. Although there are few laws regulating how high gasoline prices can be, there are bans on below-cost sales in at least 10 states. These bans set the price above a threshold cost, or some amount above the wholesale cost. These controls prevent stations from excessively undercutting their competition and driving them out of business. Thus, the FTC report concluded that gasoline is purchased and sold at a price that reflects the spot price of oil and the demand for gasoline. Consumers purchase gasoline at a price that almost immediately reflects the oil market, as otherwise a station owner loses money. After a decline in the price of oil, prices at the pump can remain high if a gas station buys or anticipates buying more gasoline at the former higher price, since the station must continue to sell gasoline at that higher price until lower-priced gasoline has been bought and sold. As one study of California gasoline prices put it, "After crude oil and wholesale gasoline prices peak and start to decline, retail prices may still be "digesting" the effects of the previous increase, even while starting to reflect the decrease as well."[77]

However, there is another and perhaps more fundamental reason for the higher price of gasoline even after the oil price has declined: the profit motive of the outlet. An outlet owner can continue to sell gasoline at the higher price, since consumers have become used to paying it. What stops the outlet owner?

At some point, competing stations lower their prices, forcing other stations to lower theirs. From a market standpoint, such pricing behavior is not gouging because the station owner is not trying to extract a price and profit higher than the market will bear. Instances of collusion or abuse are rare, and outlet owners are merely making a profit, however annoying that might be to consumers. In California, after the gasoline price increases following hurricane Katrina in 2005, there were 1,150 complaints of gouging received by the state. Of those, 50 were selected for investigation and none was found to have broken the law.[78] Gasoline price gouging did occur after hurricane Katrina, when the FTC confirmed 15 cases, but only one of those was flagrant, and the others were attributed to confusion in the marketplace.[79] What does appear to be the case is that such "excess" profit-making is temporary. Thus, the primary controls on the short-term price of gasoline are the costs of crude oil, getting the finished gasoline to market, and station owners protecting themselves from loss. The long-term price trend tracks the price of oil.

Points to Take Away

A lot of facts and figures were presented in this chapter. The following is a summary of key points about oil and its production and consumption.

Definitions

- Petroleum is oil plus natural gas.
- The oil endowment is the sum of the estimated oil resource plus all prior cumulative production. Reserves are the known resources anticipated to be produced profitably. The oil deemed technically recoverable depends on technology that may or may not become cost-effective.

Oil Company Views of Peak Oil

- There is disagreement among major oil companies on the issue of whether the world will soon face a decline in oil production due to limited global reserves.

Oil Company Profits

- During the 15 years leading up to 2000, the major US-based energy companies had a return on investment (ROI) of 2 percent less than All Manufacturing Companies, the US Census Bureau benchmark.

- From 2000 to 2008, the ROI of the major US-based energy companies was 7 percent more than the All Manufacturing Companies benchmark.
- Oil company profits have been greatest following the onset of wars and other crises.

OPEC

- The Organization of Petroleum Exporting Countries (OPEC) – Algeria, Angola, Ecuador, Iran, Iraq, Kuwait, Libya, Nigeria, Qatar, Saudi Arabia, United Arab Emirates, and Venezuela – has significantly controlled the price of oil by limiting production.
- OPEC produces over 40 percent of global oil.
- OPEC has about 70 percent of global oil reserves (56 percent is in the Middle East).

Global Oil

- The US Geological Survey estimates that the global conventional (liquid) oil endowment exceeds 3 trillion barrels, of which about one-third was produced through 2008. The Assessment horizon was the 30 years from 1996 to 2025.
- As of January 2009, there were 1.34 trillion barrels of reported global reserves, which is oil that is known to exist and can be profitably produced. At the 2008 rate of production, these reserves would provide sufficient supply for about 50 years.
- In decreasing order, Saudi Arabia, Russia, and the US produce one-third of the world's oil.
- Annual global oil consumption exceeds production by 15 percent because of gains in liquids from oil refining and natural gas processing. All oil is accounted for when balancing production and consumption.

US Oil

- The US has less than 2 percent of global oil reserves.
- The US consumes 24 percent of global oil, more than twice as much as any other country.
- In the US, annual per capita oil consumption is 25 barrels per person, versus 2 barrels per person in China and 1 barrel per person in India.
- The US imports two-thirds of its oil – half from OPEC, one-third from Canada and Mexico, and the remainder from other countries.

- Only 6 percent of US oil demand is satisfied by oil from the Persian Gulf region.

Oil Products and Quality

- About 80 percent of each barrel of oil is made into transportations fuels, primarily gasoline, diesel, and jet fuel. About half of each 42-gallon barrel becomes gasoline.
- Oil quality – light versus heavy and sweet versus sour – has affected the price of different oils by about 10 percent.

Gasoline

- About half to three-fourths of the cost of gasoline in the US is for the raw material, oil.
- Gasoline prices follow oil prices. A $10 increase (*decrease*) in the price of oil results in an increase (*decrease*) in the US average price of gasoline of 25 cents per gallon.
- Gasoline prices in the US are less than half those in Europe because two-thirds of the price Europeans pay is for taxes, versus about 40 cents (roughly 10 to 20 percent with US gasoline at $4 to $2 per gallon) for taxes in the US. Some countries subsidize the gasoline prices paid by their consumers by 80 to 90 percent.
- Over the short term, US gasoline consumption is fairly unresponsive to price increases. Consumption decreases by approximately 0.5 percent for every 10 percent increase in price. Over the longer term, consumption declines by approximately 5 to 6 percent for every 10 percent increase in price as consumers restructure their spending behavior. For small changes in price, consumption also increases in these proportions as the price of gasoline decreases.
- Retail gasoline prices can jump suddenly because station owners set prices based on the cost of their *next* delivery. Prices may be slower to come down as an owner tries to maximize profit without losing business to competing stations.

Notes and References

1. Exxon Mobil, op-ed piece, March 2006, http://exxonmobil.com/Corporate/Files/Corporate/OpEd_peakoil.pdf
2. Dukes, J. S. (2003). *"Burning Buried Sunshine: Human Consumption of Ancient Solar Energy," Climatic Change*, **61**(1–2): 31–44 (14), Springer.

3. US Geological Survey (2009). *Mineral Commodity Summaries 2009.* Washington, DC: US Government Printing Office.

4. US Minerals Management Service (2006). "Assessment of Undiscovered Technically Recoverable Oil and Gas Resources of the Nation's Outer Continental Shelf, 2006," www.mms.gov/revaldiv/PDFs/2006NationalAssessmentBrochure.pdf

5. US Geological Survey (2009). *Mineral Commodity Summaries 2009.* Washington, DC: US Government Printing Office.

6. US Minerals Management Service (2006). "Assessment of Undiscovered Technically Recoverable Oil and Gas Resources of the Nation's Outer Continental Shelf, 2006," www.mms.gov/revaldiv/PDFs/2006NationalAssessmentBrochure.pdf

7. US Minerals Management Service (2003). "Assessment of Undiscovered Technically Recoverable Oil and Gas Resources of the Nation's Outer Continental Shelf, 2003 Update," www.mms.gov/revaldiv/PDFs/2003NationalAssessmentUpdate.pdf

8. www.lloydminsterheavyoil.com/petrochem01.htm#Hydrocarbons

9. www1.eere.energy.gov/vehiclesandfuels/pdfs/deer_2002/session1/2002_deer_eberhardt.pdf

10. Yergin, D. (1991). *The Prize: The Epic Quest for Oil, Money, and Power.* New York: Free Press, Simon and Schuster.

11. *Derrick's Hand Book of Petroleum*, Volume 1. Oil City, PA: Derrick Publishing Company, 1898: 77. From: www.sizes.com/units/barrel_petr.htm and comments by Jean Laherrère on the article by P. Holberg and R. Hirsch, "Can we identify limits to worldwide energy resources," *Oil and Gas Journal*, June 30, 2003: 20–6. The origin of bbl is challenged in www.unc.edu/~rowlett/units/, which suggests that this abbreviation was already in use in the 1700s. That does not necessarily negate the history of the use of "bbl" versus "bl" for oil.

12. Energy Information Administration (2008). "Performance Profiles of Major Energy Producers 2007," www.eia.doe.gov/emeu/perfpro/020607.pdf

13. www.eia.doe.gov/emeu/perfpro/aboutCOs.pdf. In 2007, the 25 energy companies reporting to the US Department of Energy as part of their Financial Reporting System (FRS) were: Alenco Inc., Exxon Mobil Corporation, Amerada Hess Corporation, Lyondell Chemical Corporation, Anadarko Petroleum Corporation, Marathon Oil Corporation, Apache Corporation, Motiva Enterprises, LLC, BP America, Inc., Occidental Petroleum Corporation, Chesapeake Energy Corporation, Shell Oil Company, Chevron Corporation, Sunoco, Inc., CITGO Petroleum Corporation, Tesoro Petroleum Corporation, ConocoPhillips Company, The Williams Companies, Inc., Devon Energy Corporation, Total Holdings USA, Inc., El Paso Corporation, Valero Energy Corp., EOG Resources, Inc., XTO Energy, Inc., and Equitable Resources, Inc. Note that five of the FRS companies are owned by foreign companies: Alenco, owned by Encana Corporation; BP America, owned by BP plc; CITGO, owned by Petroleos de Venezuela, SA; Shell Oil, owned by Royal Dutch Shell plc; and Total Holdings USA,

owned by Total SA. In 2006, the FRS company list included Kerr-McGee Corporation, Burlington Resources, Inc., Premcor, Inc., and Dominion Resources. Burlington Resources and Kerr-McGee were acquired by ConocoPhillips and Anadarko Petroleum, respectively, in 2006. Dominion Resources sold most of its oil and natural gas production assets and no longer met the selection criteria for the FRS.

14. Of course, this $5,000 figure is given merely to provide a sense of how much money $40 trillion represents. The $5,000 is only for the current population, while the $40 trillion figure is for all oil sales since 1859.

15. www.eia.doe.gov/emeu/perfpro/dscytables2007.xls; www.eia.doe.gov/emeu/perfpro/aboutCOs.pdf

16. "Chevron's deepwater Jack #2 well touted as breakthrough GoM discovery," *Oil and Gas Financial Journal*, **3**(10), 2006. Langley, D. (2006). "Lower Tertiary Trend: A Study in the Impact of Advancing Technology," *Journal of Petroleum Technology*, **58**(12), December 2006. Mufson, S. (2006). "U.S. Oil Reserves Get a Big Boost: Chevron-Led Team Discovers Billions of Barrels in Gulf of Mexico's Deep Water," *Wall Street Journal*, December 6, 2006. www.chevron.com/news/press/Release/?id=2006-09-05

17. "Performance Profiles of Major Energy Producers 2007," www.eia.doe.gov/emeu/perfpro/tab01.htm

18. www.eia.doe.gov/emeu/perfpro/020607.pdf

19. Pirog, R. (2007). *The Role of National Oil Companies in the International Oil Market*, Congressional Research Service Report, August 21, 2007.

20. Greene, D. L. (2009). "Measuring Energy Security: Can the United States Achieve Oil Independence?" *Energy Policy* (in press: doi:10.1016/j.enpol.2009.01.041).

21. Energy Information Administration reporting of Oil and Gas Journal value (January 2008); *BP Statistical Review* (2008); www.opec.org/home/basket.aspx

22. The average price of OPEC oil in 2008 was $95 per barrel. This gave a total value of OPEC reserves of about $87 trillion. The gross world product in 2008 was $69 trillion (IMF estimate in purchasing power parity, ppp). www.imf.org/external/pubs/ft/weo/2008/01/pdf/tables.pdf

23. Reported here are data from the US Energy Information Administration. The EIA does not include in its OAPEC oil production statistics Egypt, Syria, and Bahrain (the non-OPEC members).

24. Note that Middle Eastern OPEC countries are low-cost producers at $2 per barrel. Shafiq (2009) estimated finding and development costs of Iraq's oil at $1.50–2.25 per barrel. Iraq has 8 percent of global reserves. Shafiq, T. (2009). "Iraq's oil prospects face political impediments," *Oil and Gas Journal*, January 19, 2009: 46–9. The EIA Financial Reporting System (2008) documents US-based multinational oil company finding costs and production costs per barrel in the US of $17.01 (in 2005–7) and $8.35 (in 2007, net of taxes), respectively. These companies' Middle East per barrel finding costs were $14.85 in 2005–7, and their production costs were $4.08 in 2007. Their worldwide average finding

($9.98 with taxes and $7.35 net of taxes in 2007) and lifting costs combined averaged $27.10 per barrel including taxes in 2005–7, up from $17.74 per barrel in 2003–5. www.eia.doe.gov/emeu/perfpro/020607.pdf

25. Energy Information Administration, "OPEC revenues fact sheet," March 2009.
26. Figures based on EIA data and include lease condensates but not production gains. OAPEC percentages from 1970 to 1980 were estimated based on the fraction of OAPEC to OPEC production, including processing gains and regression to 1980–90 known values.
27. EIA data for average OPEC countries average spot oil price: http://tonto.eia.doe.gov/dnav/pet/hist/wtotopecw.htm
28. Green, D. L., and P. N. Leiby (2006). "The Oil Security Metrics Model," Oak Ridge National Laboratory, ORNL/TM-2006/505.
29. EIA data for oil production, including lease condensate: OPEC oil production for the seven years from 1973 to 1979 averaged 30.2 million barrels per day (11.02 billion barrels per year), and between 2001 and 2008, it averaged 31.2 million barrels per day (11.36 billion barrels per year), a 3 percent increase. Reserves based on EIA data comparing 1980 and 2008 values. Reserve estimates for 2008 from *Oil and Gas Journal*, "Worldwide look at production and reserves" (January, 2009).
30. Mouawad, J. (2008). "OPEC Says it Will Cut Oil Production," *The New York Times*, September 9, 2008.
31. Fletcher, S. (2009). "Market Watch: Crude prices continue downward spiral," *Oil and Gas Journal*, January 13, www.ogj.com/display_article/350153/7/ONART/none/GenIn/1/MARKET-WATCH:-Crude-prices-continue-downward-spiral/
32. Mouwad, J. (2008). "OPEC Looks to Halt Falling Oil Prices," *The New York Times*, December 16, 2008.
33. "Emirates cuts oil supplies," Associated Press, published in *International Herald Tribune*, February 26, 2008, www.iht.com/articles/ap/2009/02/26/business/ML-Mideast-OPEC.php
34. "Saudi Arabia Committed to Oil Market Stability," Associated Press, *The New York Times*, January 14, 2009.
35. "OPEC: World will pay for low oil prices by 2013," *Bloomberg News*, March 6, 2009.
36. "OPEC Commentary – Dependence is a two-way street," April 2007, www.opec.org/opecna/commentaries/2007/Comm042007.htm
37. US Geological Survey World Petroleum Assessment Team (2000). *U.S. Geological Survey World Petroleum Assessment 2000—Description and Results*. US Geological Survey Digital Data Series DDS-60, 4 CD-ROMs. Also see Ahlbrandt, T. S., R. R. Charpentier, T. R. Klett, J. W. Schmoker, C. J. Schenk, and G. F. Ulmishek (2005). "Global Resource Estimates from Total Petroleum Systems," AAPG Memoir 86, American Association of Petroleum Geologists, Tulsa, OK.
38. The eight regions roughly correspond to US State Department regions.

39. The effective date of the USGS 2000 Assessment is January 1, 1996, so all oil supply and consumption data in the Assessment are for that date.

40. http://energy.cr.usgs.gov/WEcont/chaps/AR.pdf Report cites: NRG Associates, Inc. (1995) The significant oil and gas pools of Canada data base, NRG Associates, Inc., Colorado Springs, CO; and Petroconsultants (1996) Petroleum exploration and production database: Houston, Texas.

41. Horn, M. (2007). "Giant fields likely to supply 40 percent+ of world's oil and gas," *Oil and Gas Journal*, April 9, 2007: 35–7.

42. Ahlbrandt, T. S., R. R. Charpentier, T. R. Klett, J. W. Schmoker, C. J. Schenk, and G. F. Ulmishek, (2005). "Global Resource Estimates from Total Petroleum Systems," AAPG Memoir 86, American Association of Petroleum Geologists, Tulsa, OK.

43. US Geological Survey National Oil and Gas Resource Assessment Team (1995). "1995 National Assessment of United States Oil and Gas Resources," U.S. Geological Survey Circular 1118: 20 pp.

44. The US has fewer reserves than Kazakhstan, but the US is third in global oil production with half a million producing oil wells. Kazakhstan has about 700 producing wells. Collectively, the OPEC nations have 37,000 producing oil wells, less than 8 percent of the number in the US.

45. www.bp.com, data tables from *BP Statistical Review of World Energy*, subsequently cited as *BP Statistical Review*.

46. The Organization for Economic Cooperation and Development (OECD) countries are Austria, Belgium, Czech Republic, Denmark, Finland, France, Germany, Greece, Hungary, Iceland, Ireland, Italy, Luxembourg, the Netherlands, Norway, Poland, Portugal, Slovakia, Spain, Sweden, Switzerland, Turkey, and the United Kingdom.

47. China customs information for 2006, reported by the China Institute at the University of Alberta, courtesy of Simin Yu, 2007.

48. EIA data.

49. *BP Statistical Review*.

50. Andrews-Speed, P. (2006). "China's energy and environmental policies and their implications for OPEC," *The CEPMLP Internet Journal*, **17**(6); Centre for Energy, Petroleum and Mineral Law and Policy, University of Dundee, citing State Bureau of Customs of China, cited in *International Petroleum Economics*, **13**(3), 2005, and in *International Petroleum Economics*, **9**(3), 2001.

51. Analysis of 2007 data from the China Institute at the University of Alberta suggests annual imports of 543 million barrels from Saudi Arabia, Angola, Iran, and Venezuela, which would indicate that at least 43.3 percent of imports and 20.5 percent of China's demand were derived from OPEC nations.

52. Angola, which became an OPEC member in 2007, provided 16 percent of China's imported oil in 2006. If Angola had been counted as part of OPEC in 2006, OPEC's portion of China's imports would have been over 57 percent, and OPEC would have supplied 24 percent of China's oil demand. Another Persian Gulf country, Oman, supplied China with another 9 percent of its imports.

53. Energy Information Administration, 2007, taken as the average April 2006–March 2007 refinery products, http://tonto.eia.doe.gov/dnav/pet/pet_pnp_pct_dc_nus_pct_m.htm
54. EIA and www.energy.ca.gov/gasoline/whats_in_barrel_oil.html
55. Natural gas plant liquids do not include lease condensate.
56. Analytical Testing Services, Inc., Franklin, PA, www.wetestit.com
57. At concentrations of 500–700 ppm, loss of consciousness in 30 to 60 minutes; at 1,000–2,000 ppm, death within minutes. From www.michigan.gov/deq
58. Society of Petroleum Engineers, Glossary: www.spe.org/spe-app/spe/industry/reference/glossary.htm
59. Prior to June 2005, the OPEC Basket mirrored seven crude oils from six OPEC countries – Algeria, Dubai, Indonesia, Nigeria, Saudi Arabia, and Venezeula – and one non-OPEC country, Mexico. Its average sulfur content was 1.44 percent. (EIA, 2008, "OPEC Brief," www.eia.doe.gov/cabs/opec.html, and OPEC.org)
60. The range since 2001 in these premium percentages has been significant: 7 percent to 15 percent for WTI vs. OPEC Basket, and 3 percent to 8 percent for Brent vs. OPEC Basket.
61. 2005 Statement of John Cook, Director, Petroleum Division, Energy Information Administration, US Department of Energy before the Subcommittee on Energy and Resources Committee on Government Reform, US House of Representatives, May 9, 2005, and EIA data.
62. Notes: Data based on Department of Energy Financial Reporting System information (December, 2008). All values are in 2007$. Lifting costs include production taxes and are values in Performance Profiles of Major Energy Producers. The decline in lifting cost from 2006 to 2007 in the Middle East primarily was due to a reduction in production taxes from $10.21 to $4.52 per barrel. Finding costs for 2006 and 2007 are 2004–6 and 2005–7 reporting period values. Finding cost values for the former Soviet Union (FSU) were estimated based on EIA FRS data assuming costs increased as in Western Europe. A lower estimate of $15.74 is based on the increase in FSU lifting costs. www.eia.doe.gov/emeu/perfpro/t&f_o&g.html
63. Inflation adjustments based on CPI given by R. Sahr, http://oregonstate.edu/cla/polisci/faculty-research/sahr/sahr.htm
64. Hargreaves, S. (2008). "U.S. Gas: So cheap it hurts," *CNN Money*, July 15, 2008; Bello, D. (2008). "The Price of Gas in China," *Scientific American*, August 4, 2008.
65. China raised gasoline prices by 20 percent in June 2008; the value shown is estimated based on the price relative to that in the US in early August 2008.
66. Note that in economic terminology, demand is price elastic if the price elasticity of demand value is greater than 1 and inelastic if the value is less than 1. For example, if the percent decrease in demand is 2.2 percent given a 10 percent price increase, the price elasticity of demand is 0.22. This low price elasticity, or price inelasticity, indicates that price changes of the commodity have

relatively little effect on consumption behavior. A small price change is assumed when determining elasticity values. In these terms, short-run gasoline price elasticity is 0.05, and the long-run value is 0.60; both are price inelastic.

67. Hughes, J. E., C. R. Knittel, and D. Sperling (2008). "Evidence of a Shift in the Short-Run Price Elasticity of Gasoline Demand," *The Energy Journal*, **29**(1).

68. Cooper, J. C. B. (2003). "Price elasticity of demand for crude oil: estimates for 23 countries," *OPEC Review*, **27**: 1–8.

69. US Federal Trade Commission (2005). "Gasoline price changes: The dynamics of supply, demand, and competition," page 135, citing Epsey, M. (1998). "Gasoline Demand Revisited: An International Meta-Analysis of Elasticities," *Energy Economics*, **20**: 273–95, for long-term elasticity value and their own analysis for the short-term value.

70. Cooper, J. C. B. (2003). Price elasticity of demand for crude oil: estimates for 23 countries, *OPEC Review*, **27**: 1–8.

71. www.energy.ca.gov/gasoline/gasoline_q-and-a.html

72. Ye, M., J. Zyren, J. Shore, and M. Burdette (2005). "Regional Comparisons, Spatial Aggregation, and Asymmetry of Price Pass-Through in U.S. Gasoline Markets," *Atlantic Economic Journal*, **33**: 179–92.

73. Energy Information Administration (2003). *California Gasoline Price Study Final Report*, November 2003: 49.

74. "House passes gasoline gouging bill," by Richard Simon, *Los Angeles Times*, May 23, 2007.

75. US Federal Trade Commission (2005). "Gasoline price changes: The dynamics of supply, demand, and competition," page 135.

76. Gouging would only be profitable for a commodity whose price elasticity of demand is relatively inelastic, like gasoline. Otherwise, raising the price would result in a loss of revenue because demand would decrease significantly.

77. Energy Information Administration (2003). *California Gasoline Price Study Final Report*, November 2003: 48.

78. "Anti-gouging laws don't cut gas prices. State probed 50 potential cases; no charges," by D. R. Baker, *San Francisco Chronicle*, May 10, 2006.

79. "FTC Head Opposes Anti-Gouging Law; Says Regulation Would be Hard to Enforce and Could Cause Fuel Shortages," CBS/AP, May 23, 2006, www.cbsnews.com/stories/2006/05/22/business/main1639514.shtml

3

The Historical Resource Depletion Debate

Global oil depletion can be viewed in the context of long-standing concerns that humans have relied so heavily on Earth's natural resources that we are surely going to run out of some essential natural commodities. There are classical historical arguments that have been used to support the notion of natural-resource exhaustion and the scientific basis for the decline in global oil production. The oil depletion debate has occurred in the context of repeated panics that the world is entering an oil-supply crisis. Has the crisis finally arrived?

The Malthusian Doctrine

In 1798, Thomas Malthus, a 32-year-old British economist and demographer, published *An Essay on the Principle of Population as it Affects the Future Improvement of Society*. There, he argued that society as it was then known was not sustainable:

> ... I say, that the power of population is indefinitely greater than the power in the earth to produce subsistence for man. ... Population, when unchecked, increases in a geometrical ratio. Subsistence increases only in an arithmetical ratio. ... I see no way by which man can escape from the weight of this law which pervades all animated nature.[1]

Malthus concludes:

> Famine seems to be the last, the most dreadful resource of nature. The power of population is so superior to the power in the earth to produce subsistence

Oil Panic and the Global Crisis: Predictions and Myths. 1st edition. By Steven M. Gorelick.
Published 2010 by Blackwell Publishing, ISBN 978-1-4051-9548-5 (hb)

for man, that premature death must in some shape or other visit the human race.[2]

Malthus believed that population would continue to grow exponentially in a positive feedback system, with growth spawning further growth, and would outpace the world's ability to produce food, which he thought would at best grow linearly (e.g., in proportion to the amount of cultivated arable land). Although regional famines have been devastating and continue to occur, Malthus did not foresee the technological advances in agriculture and food production that have accompanied the increasing global population. Malthus was the first apocalyptic voice for the inevitability of resource depletion, and he was wrong.[3]

Malthus engaged in a type of blind projection. Had the population of his day (980 million) doubled every 25 years, as he suggested it might, the world population would be over 300 billion in 2009 (rather than the actual 6.8 billion in 2009).[4] Yes, there are food shortages and devastation accompanying natural disasters and social turmoil, but we can produce enough food to supply the world. We have the means to prevent massive starvation, even if we don't.[5]

The Limits to Growth

Neo-Malthusian doctrine has dominated the resource depletion debate. In 1972, a group calling itself the Club of Rome and led by researchers at the Massachusetts Institute of Technology (MIT) produced *The Limits to Growth*, a report on the predicament of mankind. This group of scholars undertook an ambitious project of developing a computer model of the world – a virtual world system whose response to assumptions about resource availability and public policies about pollution and population controls could be explored. Their model recognized the compounding effect of exponential growth in population on the depletion of natural resources, but it attempted to counterbalance this effect by including potential negative feedback loops, such as decreased life expectancy caused by low nutritional levels or greater pollution. Covering the period from 1900 through 2100, their model was used to predict population, natural resources, industrial output, food consumption, and pollution (including soil erosion). Under a variety of scenarios, they predicted population overreaching available resources followed by rapid collapse in food availability, industrial output, and indeed human population itself. In fact, assuming no changes in behavior from 1900 through 1970, their historical period for which they analyzed global data, they concluded, "We can thus say with some confidence that, under the assumption of no major change in the present system, population and industrial growth will certainly stop within the next century, at the latest."[6] The century referred to is the twenty-first century, and their predictions are shown in Figure 3.1. They called the population decline phenomenon "overshoot and collapse."

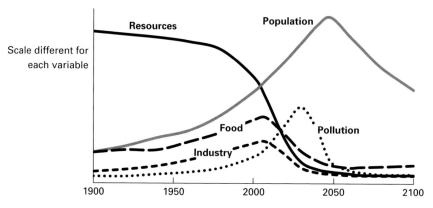

Figure 3.1   Status quo (business as usual) world model predictions from the Club of Rome study showing overshoot and collapse of society as the global carrying capacity is exceeded. Each variable is plotted on its own scale and each curve reflects relative changes. Industrial output and food are per capita (after Meadows et al. (1972), *The Limits to Growth*,[7] with curves rescaled).

The *"Limits to Growth"* model was almost immune to factors that one might guess would prevent overshoot and collapse. Using any reasonable assumption for the future resource base or public policies aimed at restraint, the world model still predicted disintegration of global systems. For example, when the resource base was doubled, industrial production increased to a higher level, but global pollution caused an increase in the death rate, a decline in food production, and population collapse. When it was assumed that, starting in 1975, pollution generation per unit of industrial and agricultural output was only a quarter of 1970 values and resources were unlimited, arable land reached its limit of production, resulting in food shortages and consequently the crash of global population. Adding birth control to the model only delayed the population collapse by a decade or two. Even recycling of raw materials did not prevent the seemingly unavoidable increase in industrial growth, the pollution it generated, and eventual population decline.

The model predicted a stable world system only if simultaneous draconian measures were taken: the birth rate fixed at the 1975 death rate, global resource consumption restricted to its 1970 value, industrial production replaced by services such as health and education, pollution per unit of industrial and agricultural production set at a quarter of its 1970 value, global food supply distributed equitably even if it was uneconomic to do so, soils enriched through urban composting, and greatly diminished obsolescence of industrial equipment. With these fantastic assumptions, a stable world population and economy were shown to be possible. Under this scenario, the world's predicted stable industrial output would be three times that of 1970.

The neo-Malthusian *Limits to Growth* model adopted an underlying principle of ecology, "carrying capacity." According to this principle, in an environment or an entire world with finite resources, population has limits. Just as animals without natural predators can overpopulate, overgraze, and cause soil erosion that leads to permanent loss of vegetation, resulting in collapse of the animal population, the *Limits to Growth* study suggested that, globally, the human population might exceed Earth's carrying capacity. However, just as Malthus had it wrong, so too did the Club of Rome. Their model failed to recognize that human ingenuity gives rise to new technology for industry and agriculture, and this innovation has outpaced the demands of increased population and longer life spans. The Club of Rome's dire prediction of a world system near collapse failed to materialize. No fundamental scarcity of any global non-renewable resource, including oil, has occurred, and the forecasting approach of the world model was vitally flawed.

The neo-Malthusian position has persisted with Hubbert's conjecture of global oil depletion. There have been predictions by some oil analysts of chaos and perhaps a major transformation of civilization as we know it owing to over-dependence on oil, which many believe will reach peak production very soon and decline substantially within the next few decades. The imminent depletion of global oil resources has been predicted in published works many times since oil was discovered. Examples are:

- "It is concluded that the critical date per USGS data when global oil demand will exceed the world's production will fall somewhere between 2000 and 2010, and may occur very suddenly due to unpredictable political events. This is within the lifetimes of most people now alive. This foreseeable energy crisis will affect everyone on earth."[8]
- "We should not forget the universal laws that 'What goes up must come down' and 'What is born shall die'. These are true for human beings, their civilization, and even the Earth itself."[9]
- "... the point beyond which oil production will irreversibly start declining is approaching and will be reached, even according to the most optimistic scenarios, before 2040 and quite possibly much sooner. If the appropriate solutions for substituting crude oil and for conserving the use of energy

are not implemented in time, then the current upward trend in oil prices is bound to continue."[10]

- "He [Campbell] forecasts, 'The World will become a very different place with a smaller population. The transition will be difficult, and for some catastrophic' with 'major economic and political discontinuity' globally and 'great suffering' (Campbell, 1997, p. 177)."[11]
- "Intervention by governments will be required, because the economic and social implications of oil peaking would otherwise be chaotic. … The world has never faced a problem like this. Without massive mitigation more than a decade before the fact, the problem will be pervasive and will not be temporary. Previous energy transitions (wood to coal and coal to oil) were gradual and evolutionary; oil peaking will be abrupt and revolutionary. … Without mitigation, the peaking of world oil production will almost certainly cause major economic upheaval."[12]

The Oil Panics of 1916 and 1918

One of the first major oil depletion scares occurred near the turn of the last century. The Ford Model T automobile was a novelty in 1908, with sales of just 10,000 cars. By 1914, sales were up to 200,000, with gasoline selling for 16 cents per gallon. Oil was plentiful, and a 1914 oil glut generated a price drop from $1.05 to $0.55 per barrel. But in July, World War I began. Several years before, in 1911, the US Navy, with some foresight, initiated production of nimbler US battleships powered by oil rather than coal. Similarly, in 1913, a young Winston Churchill, then First Lord of the Admiralty, began the transformation of British battleships to run on oil. However, as World War I got underway, there was a collision between over-demand and under-supply.

In 1915, the demand for fuel made from oil for both military and merchant marine ships was coupled with the need for gasoline for the 2.1 million cars in the US. The supply situation also changed that year, with the unexpected decline in yield from Oklahoma's Cushing oil field, then the most productive region in the world. In 1916, a crisis unfolded, with Oklahoma oil prices jumping from a low of 40 cents a barrel in 1915 to as much as $2.05 a barrel in early 1916.[13] Predictions of doom followed. The US Bureau of Mines reported to the US Senate in 1916 that peak production would occur within five years, and stated,[14] "In the exhaustion of its oil lands and with no assured sources of domestic supply in sight, the United States is confronted with a national crisis of the first magnitude." The Director of the Bureau, Dr. Van H. Manning, predicted exhaustion of US supplies within 27 years.[15]

The crisis came to an end later that same year when two things occurred: the new technology of "cracking" oil to make gasoline was adopted by oil refiners, and high prices spawned more drilling with consequent discoveries in both California and the mid-continent. By year's end, oil prices had fallen from $2.05 to 90 cents per barrel.

Serenity was short-lived. By mid-1918, there were almost 5.5 million motor vehicles in the US, and the war continued to require that oil be shipped overseas. The US produced only 400 aircraft in 1916, yet that number grew to over 11,000 in 1918.[16] The excessive fuel demand drove US Oil Administration Secretary Requa to appeal to patriotic citizens to abide by gasless Sundays. Oil exhaustion was again the word of the day.[17] Gilbert and Pogue of the Smithsonian Institute in 1918 claimed,[18] "There is no hope that new fields, unaccounted in our inventory, may be discovered of sufficient magnitude to modify seriously the estimate … [The war] has merely brought into the immediate present an issue underway and scheduled to arrive in the course of a few years." The crisis continued into 1919 with the dire prediction by David White, USGS Chief Geologist, that, "… the peak of [US] production will soon be passed – possibly within three years."[19]

However, as the war ended in November 1918, demand began to subside, and at the same time, increasing new oil production in Texas helped increase US supplies to more than 1 million barrels per day in the following year.

The threat that oil would become scarce became a serious issue again and again during the remainder of the twentieth century. Here is one notable quote just for the period before the US entered into World War II:

> … it is unsafe to rest in the assurance that plenty of petroleum will be found in the future merely because it has been in the past. (L. Snider and B. Brooks, *American Association of Petroleum Geologists Bulletin*, 1936[20])

Panic Revisited: The Oil Crisis of the 1970s

In 1973, OPEC twice raised the price of oil. The first price increase of 70 percent was a simple show of muscle. The second increase was tied to the Arab–Israeli war, which began as Egyptian and Syrian forces invaded Israel on the 1973 Yom Kippur holiday. Various countries supported Israel, including the US, the Netherlands, Portugal, South Africa, and Rhodesia. Outraged, OAPEC (the Organization of Arab Petroleum Exporting Countries) established an embargo, refusing to sell oil to these countries. They cut supply 5 percent per month to unfriendly nations (i.e., nations friendly to Israel). In total, OAPEC cut production from about 21 to 16 million barrels per day (a total of over 23 percent). Production quickly increased in the rest of the world, and the net result was only a 9 percent reduction in global supply.[21] This seemingly small reduction combined with great uncertainty resulted in a dramatic price increase. Saudi light crude went from $1.90 per barrel in 1972 to $9.60 in 1974.[22] There was panic in the marketplace. US filling stations ran out of gasoline. The US retail price of gasoline went up by 40 percent in 1973, there were long lines of cars waiting to refuel, and sales were often restricted to every-other-day based on even and odd license plate numbers.[23]

Although OAPEC's embargo was lifted in March 1974, oil production remained tight, and prices continued to climb. Was the world running out of oil? President Jimmy Carter thought so and stated in 1977 that "We could use up all the proven reserves in the entire world by the end of the next decade. …"[24] At the same time, the National Energy Program, Executive Office of the President, decreed:

> The diagnosis of the U.S. energy crisis is quite simple: demand for energy is increasing, while supplies of oil and natural gas are diminishing. Unless the U.S. makes a timely adjustment before world oil becomes very scarce and very expensive in the 1980s, the nation's economic security and the American way of life will be gravely endangered.[25]

In 1979, the regime of the Shah of Iran was toppled, and Iran's oil exports ceased for about nine months. At the time, Iran was the world's second-largest oil exporter, and the shortage caused prices to climb again. The price of Iranian light crude went from $13 to $30 per barrel in 1980, and Saudi light crude went from $12 per barrel in 1977 to $26 over the same time-frame. Marketplace panic was re-ignited.

The world oil situation prompted the CIA to report in 1980, "We believe that world oil production is probably at or near its peak … Simply put, the expected decline in oil production is the result of a rapid exhausting of accessible deposits of conventional crude oil."[26] By 1982, Saudi light crude had reached $34 per barrel, almost three times its price just five years earlier. During the period of bloated oil prices, world economies suffered from the inflationary ripple effect. From 1972 to 1980, US consumer prices for goods and services doubled.[27]

Throughout the 1970s and into the very early 80s, the price of oil increased and created great hardship and uncertainty, but those who concluded that the world was running out of oil were mistaken. During the mid-80s, there was an oil glut, and by 1989, Saudi light crude was selling at $13 per barrel. Government-directed conservation and new oil discoveries were responsible for the glut. In the US, federal car fuel economy standards instituted in 1975 mandated that over the next decade, the mileage of new cars had to be increased from an average of 13.5 to 27.5 miles per gallon (mpg). Discoveries in Alaska, Mexico, and the North Sea added to global reserves. With the completion of the Trans-Alaska Pipeline in 1977, Alaskan oil began to flow to the rest of the US, reaching over 17 percent of US production in 1980.[28] The "oil crisis" of the 1970s, with its high oil prices and expected global depletion, simply evaporated.

Arguments Supporting Global Oil Depletion

Are the current concerns about global oil depletion raised in such publications as the 2007 GAO report, the 2005 Hirsch Report to the Department of Energy, and the 2007 piece in *Science* just false alarms?[29] What facts support the claim by modern neo-Malthusians that globally we are running out of oil?

Oil analysts have built upon the work of Hubbert and used many different types of data to support the case for depletion of global resources. They make several arguments about why peak oil and a world oil production decline are upon us.

Declining Oil Production in Countries in Addition to That in the US

The US is the largest oil-producing nation that has experienced peak oil production, but other countries also have followed a pattern of decline (Figure 3.2). British Petroleum reports that there are at least 25 countries producing oil below their peak values by 20 percent or more. This reduction is the

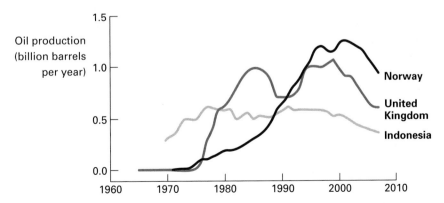

Figure 3.2 Examples of oil production declines for countries besides the US. (Data: *BP Statistical Review*; EIA)

equivalent of over one-fifth of 2008 global production. US Department of Energy data show 101 countries that have produced oil since 1980. Of these, 33 produced at a 10-year low after 2005. Although most of these 33 countries are not significant oil producers, five of them account for almost 15 percent of global production, and four of them – the US, Norway, the UK, and Indonesia – are among the world's top 20 oil producers. Compared with peak production in each country, these latter four collectively have seen a nearly 40 percent decline (more that half of this is due to the US production decline).[30]

Production Exceeds Discoveries

One worrisome trend is the mounting oil-discovery deficit, the difference between global oil discoveries and oil production (Figure 3.3). From 2000 to 2006, 85 billion barrels of recoverable oil were discovered in 140 new oil fields that ranged in size from 11.6 billion barrels down to 50 million barrels. During the same seven-year period, production totaled about 180 billion barrels. These figures suggest that the world has recently discovered only slightly less than half as much new oil as it has produced for consumption (85 versus 180 billion barrels). Furthermore, the total production potential from these recent discoveries is estimated to be merely 5.5 billion barrels per year, or just one-fifth of current global oil production (about 27 billion barrels per year in 2008).[31,32] During the early 1960s, discoveries outpaced production by nearly 10 to 1. Since the early 1980s, production has generally exceeded discoveries worldwide. The average yearly discovery-to-production ratio was about 0.67 from 1981 to 2007. That is, during the past two and a

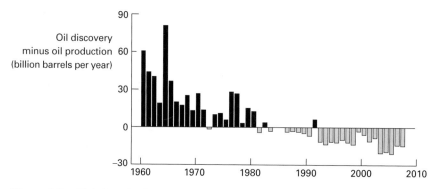

Figure 3.3 Global production has exceeded discoveries. (Data: discovery data from EnergyFiles Ltd, 2008,[33] courtesy of Michael Smith; production data from EIA)

half decades, on average the world has been running a one-third discovery-deficit relative to oil production.

Reserve and Endowment Estimates are Inflated

Followers of Hubbert fear that scarcity will begin within the next decade because estimates of global reserves are highly exaggerated, and production will drop off as reservoirs are depleted. Campbell and Laherrère (1998) state that "conventional wisdom erroneously assumes that the last bucket of oil can be pumped from the ground just as quickly as the barrels of oil gushing from wells today. In fact, the rate at which any well – or any country – can produce oil always rises to a maximum and then, when about half the oil is gone, begins falling gradually back to zero. From an economic perspective, when the world runs completely out of oil is thus not directly relevant: what matters is when production begins to taper off. Beyond that point, prices will rise unless demand declines commensurately."[34]

One problem with international reserve statistics is that they are self-reported. Estimates of reserves can and have been manipulated and may overstate the amount of readily recoverable oil. In general, such estimates are uncertain for two reasons. First, geologists and geophysicists have historically been unable to provide tight estimates of reserves for many oil fields, and estimates can range over a factor of three or more. Second, values reported by oil companies are subject to significant discretion: for example, overestimates can enhance a company's competitive advantage or support its stock price. Furthermore, oil-producing countries have their own motivations for claiming exaggerated reserve values. The best example of this motivation is the reporting by OPEC nations. Because each OPEC nation's export quota is based on the amount of its reserves, the higher the reported reserves, the higher the export quota. As displayed in Figures 3.4 and 3.5, circumstantial evidence of manipulation of reserve estimates occurred from 1988 to 1990, when more than half of the OPEC nations (Saudi Arabia, Iran, United Arab Emirates, Iraq, Kuwait, and Venezuela) increased their collective reported reserves by 77 percent, or by 304 billion barrels. In addition, with the exception of Iran, none of these countries has materially revised its reserve estimate since 1990.[35] There are, of course, other explanations for why OPEC member nations have not revised their reserve estimates in the past 10 to 20 years: they might have replaced their production with newly discovered oil; they might, as many believe, want to guard their information as a state secret; or they might just be delinquent in updating their estimates because they see no benefit in doing so.

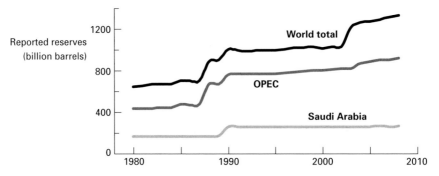

Figure 3.4   Sudden increases in reported reserves from 1988–90 were due to surprising OPEC revisions. The increase in the world total in 2003 was due to the inclusion of Canadian oil sands as part of global reserves. (Data: EIA)

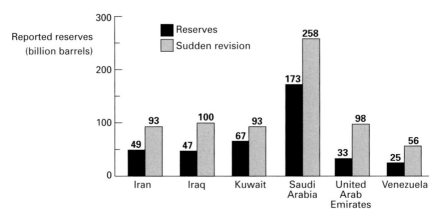

Figure 3.5   OPEC member nations showing sudden revisions of their oil reserves in the late 1980s. (Data: EIA)

The more serious concern of the followers of Hubbert – the one that can affect the projected time of "peak oil" – is that the oil endowment figures estimated by the US Geological Survey (USGS) are too high. As discussed in the previous chapter, the USGS conducted a comprehensive analysis of the global oil endowment in their 2000 *World Petroleum Assessment*. Hubbert, who died in 1989, never speculated that the oil endowment might be as high as that estimated by the USGS in their Assessment. The critics maintain that the growth of oil reserves estimated by the USGS was based on analysis of the US experience and that this model does not apply to the rest of the world. Assuming that the USGS work is technically sound, one can readily evaluate the impact of a high (3 trillion barrel) oil endowment figure on the timing of peak oil using Hubbert's method. One such projection is displayed in Figure 3.6, which estimates global peak oil production at 35 billion barrels per year in 2027. This projection fits the production data of the past decade but ignores the mismatch from 1960 to 1995. During part of that period, global oil production exceeded the estimate given by a curve based on Hubbert's method. However, using the USGS figure as the global oil endowment, this projection suggests that peak oil is less than 20 years away.[36]

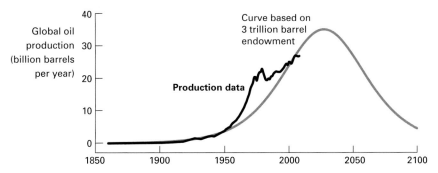

Figure 3.6   One projection of global oil production assuming a 3 trillion barrel oil endowment. (Data: production data from EIA and Hubbert (1969)[37])

Industry Exaggeration of Reserves

On January 9, 2004, Royal Dutch Shell announced the first of a succession of downward revisions of its oil reserves. In total, five revisions were announced that reduced Shell's proved reserves booked with the US Securities and Exchange Commission (SEC) by 4.5 billion barrels, or approximately 23 percent of its booked reserves in 2002. The majority of the revisions were associated with undeveloped oil regions, such as Nigeria and Oman, where oil has been produced (2.2 billion barrels), and undeveloped oil frontier regions, such as Australia, Norway, and Kazakhstan, where production is anticipated (1.2 billion barrels). Given the negative effect on shareholders, the SEC threatened suit, but in June 2005, Shell agreed to pay an SEC penalty of $120 million, and the SEC dropped its case. How could this much oil simply disappear?[38] Such revisions suggest that at least one major oil company has exaggerated its oil reserves and that the true value is much lower.

Fewer Giant Fields Discovered and Production is Declining

Most of the giant oil fields – defined as containing greater than 0.5 billion barrels – in the world have already been discovered, and future exploration is not likely to yield as many finds similar to those of the past. That is, we are discovering few new big oil fields and are relying on old and nearly spent ones for supply.

The number of giant oil fields and volume of oil in giant fields discovered during each decade since oil was first produced are shown in Figures 3.7 and 3.8, respectively.[39] The volume of oil in these giants before they were produced was estimated to be 1.4 trillion barrels[40], well over one-third of the global endowment. There are about 700 giant oil fields, each initially with at least 0.5 billion barrels of ultimately recoverable oil, discovered through 2008. Together, these giant fields account for about two-thirds of world production,[41] and most were discovered 30, 40, or more years ago. The world gets much of its oil from very few of these fields. In fact, 116 giant oil fields, each producing more than 100,000 barrels per day, are responsible for about half of global production, with 112 of them discovered more than 25 years ago.[42] The 14 largest fields are the source of one-fifth of the world's oil.[43] On average, since 2000, there have been 4.9 giant oil fields discovered per year and a total of 10.9 giant oil and natural gas fields discovered per year.[44]

Discovery of giant oil fields peaked in the 1970s and has clearly declined since then (Figure 3.7). The peak in volume of discovered oil in giant fields

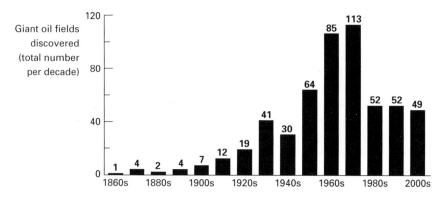

Figure 3.7   Number of giant oil fields, each containing over 0.5 billion barrels, discovered per decade, showing a marked decrease beginning in the 1980s. Note that the decade 2000 to 2010 was extrapolated based on data through 2008 (44 discoveries in total) indicating 4.9 discoveries per year. (Data: Horn (2009)[45])

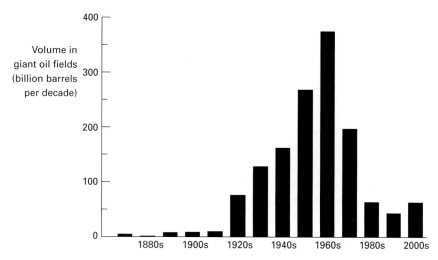

Figure 3.8 Oil volume over each decade of discovery in giant fields, including oil in giant oil and gas fields (not including natural gas condensate). The peak in global oil volume in new giant fields occurred in the 1960s and the decline has been rapid. Volume extrapolated for 2009 based on 2000–8 data. (Data: Horn (2009)[46])

occurred before the peak in the number of new oil fields (Figure 3.8), which reflects the fact that the largest oil fields were discovered first. The total volume of oil in giant oil fields discovered in the three decades since 1970 is lower than that discovered in the 1960s alone.

Production from a number of giant fields has also declined. Some 20 years ago, there were 15 oil fields producing more than 1 million barrels per day. By 2005, there were only four such fields, and all were discovered more than 30 years ago: Ghawar (Saudi Arabia), 1948; Kirkuk (Iraq), 1938; Burgan Greater (Kuwait), 1927; Cantarell (Mexico), 1976.[47]

One of the major concerns about global oil depletion has been the aging and questionable production potential of global giants, particularly in the number 1 oil-producing nation of Saudi Arabia. Much unease has centered around Ghawar, the largest oil field in the world, accounting for approximately 60 percent of Saudi oil. To maintain production at Ghawar, water is injected around the perimeter of the field to sweep the oil toward the central production wells. This practice in itself is fine, except when too much of the injected water eventually makes its way to the oil production wells. When this occurs, water as well as oil is produced. However, this in itself is not a tremendous problem as long as the **water cut** (water fraction of total liquids produced) is controlled.[48] At Ghawar, the water cut rose steadily from about

25 percent to over 36 percent in just six years (1993–9).[49] The question raised is whether Saudi Arabian oil, 90 percent of which comes from just five giant fields, can be sustained and whether or not Saudi Arabia can remain the **swing producer** (supplier with ready extra production capacity) in times of critical demand.[50]

The fact that Saudi Arabia has not changed its reserve estimate since 1990 has reinforced suspicion that the Saudis are not the oil production powerhouse they purport to be. Saudi Arabia produced 58 billion barrels from 1990 to 2008, and yet their reported reserves have remained almost constant at 262 to 266 billion barrels. The implication is that much of their **proven reserves**, those having a reasonable certainty of being extracted at a profit, appear to have been significantly depleted. For the Saudi giant fields, the constant reserve figure is now maintained by relying on what are called **possible and probable reserves**; oil for which there is either some or considerable doubt about the technical or financial viability of extraction.

There is significant concern that production from many older giant oil fields is declining. Researchers have analyzed a large database of giant fields compiled at Uppsala University in Sweden.[51,52] The database included 331 giant oil fields, of which two-thirds were onshore and one-third offshore. All told, these fields have an estimated ultimate recovery of 1.13 trillion barrels, but, by 2005, production from 79 percent of them (261 fields) was in decline at a production-weighted rate of 6 percent annually, based on an analysis of historical production. Assuming a constant rate of decline, one would expect the production from a declining field to drop by half every 11 years. The decline rate based on the Uppsala database is consistent with a study conducted by the International Energy Agency (2008),[53] which suggests an annual decline rate of 6.7 percent based on production-weighted analysis of 317 giants. Giant fields currently in production arrived at peak production at different times. Forecasts of total production from the existing global giants suggest that, collectively, they will produce at only half of their 2007 rate by about 2025.[54] However, even with declining production from existing giants, the IEA study predicts that global production will increase by about 7 percent by 2030 from the total of existing and new fields.

Decline in Discovery and Oil Drilling Suggests Onset of Production Decline

In support of the notion of peak oil and the impending decline in worldwide production, some have pointed to the similarity between the pattern of US oil discoveries and US production history. Previous oil discoveries are responsible for the oil produced today. If discoveries have declined, it makes sense that a production decline will follow. Figure 3.9 shows the Hubbert curve following the trend in US production in the coterminous US. The figure also shows historical oil discoveries following a trend similar to the Hubbert curve, but shifted back by 35 years.[55] The volume of oil in new discoveries peaked in about 1935 and has declined to just 2 percent of the peak value. If the production trend continues to mirror the discovery trend, then oil reserves in the coterminous US soon will be depleted.

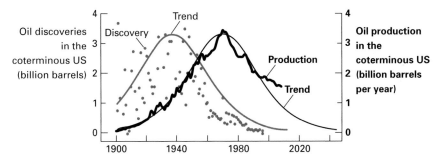

Figure 3.9   Peak in 1935 of oil discoveries in the US lower 48 states and corresponding peak in US lower 48 state oil production in 1970. Note: a similar analysis and figure is presented by Laherrère.[56] (Data: production, EIA; discovery, Klett (2003)[57])

If this staggered relation of discovery to production based on the US experience is applied to the trend in global production, peak oil production should occur some years after the peak in global discoveries. In the case of global oil depletion, the peak in production might not follow as quickly as in the US, since oil reservoirs in various locations around the world could have different life expectancies. There are about 4,000 non-giant oil fields responsible for much of the global supply, so there is a broader distribution of discovery ages in the world compared with the US. However, the total number of discoveries globally peaked in the 1980s, and the volume of discoveries peaked in the 1960s.[58] Figure 3.10 shows one possibility in which a Hubbert curve is fit to global oil production data assuming a global oil endowment of 3 trillion barrels (consistent with the USGS 2000 Assessment). In this projection, the peak in global discovery volume, which occurred in the early 1960s, is followed by possible peak production in 2027. If the trend in global oil discoveries is mirrored by that in global production, then a decline in production should be anticipated.

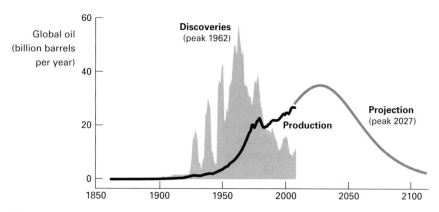

Figure 3.10   Global oil discovery volume and production data with one possible projection (gray line) peaking in 2027 (assuming 3 trillion barrels global oil endowment). Oil discoveries shown total 2.1 trillion barrels. (Data: discovery data, courtesy of EnergyFiles Ltd[59]; production data, EIA)

Global Industrial Development and Oil Consumption

A major concern of those believing that the Oil Era will soon end is the rapid increase in energy used by developing nations. Figure 3.11 shows oil use per capita versus gross domestic product (GDP) per capita, which is taken to be a measure of industrial development and average income. The points in the lower left portion of the graph correspond to poorer nations with very low per capita GDP and per capita oil consumption. These countries include India and China. Toward the upper right side of the graph are nations with higher per capita GDP and higher per capita oil consumption. Those are countries in Western Europe, Scandinavia, North America, and the developed nations rimming the Pacific. There is a clear correlation between increasing per capita GDP and increasing per capita oil consumption.

Oil-producing countries in the Middle East, such as Saudi Arabia, Qatar and the United Arab Emirates, not shown on the figure, have per capita consumption beyond the levels of the rest of the world because it is inexpensive for them to do so. Overall, these countries contribute disproportionately to total global oil consumption. For example, all OPEC nations consumed about 10 percent of the world's total oil consumption in 2007.

China and India, with low per capita GDPs, consumed less than two barrels of oil per person annually in 2007, while the relatively wealthy US consumed about 25. The concern is that when the economies of countries like China and India fully develop, their oil consumption per capita will become equivalent to that of the present-day United States. Given their enormous and

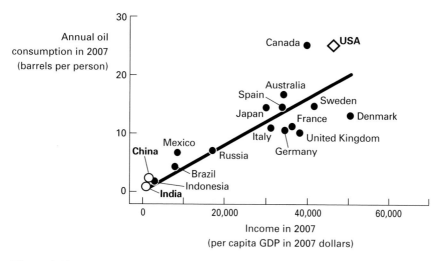

Figure 3.11 Oil consumption per capita versus average income given as GDP per capita in 2007 for selected countries. (Data: EIA and USDA Economic Research Service International Macroeconomic Data, courtesy of M. Shane)

growing populations, China and India will generate a tremendous demand for oil. After all, China alone is home to 1.3 billion people, or almost one-fifth of the world's population. It adds over 8 million people per year to its numbers and until mid-2008 had an annual industrial growth rate in excess of 10 percent.[60] Although China's industrial growth declined dramatically with the global economic crisis that began in 2008, its economy continues to grow at an enviable pace. China's economic growth has slowed but its industrial development continues to rely on oil.

China is second only to the US in its total oil use, with annual oil consumption having doubled since 1996. Its oil use has accelerated since 1980, and its oil production has not kept pace with China's growing demand (Figure 3.12). China, an oil exporter in the early 1990s, has become a significant oil importer – 1.2 billion barrels per year, which equals 4 percent of the world's annual total production. China produces less than half the oil it consumes. India's annual oil consumption has doubled in the past 14 years (Figure 3.13) and now exceeds 1 billion barrels per year. It produces less than one-third of what it consumes. With increasing income and development, it appears likely that these developing nations could eventually consume much of the world's oil. Imagine if the China of the future consumed not 2 barrels per capita annually but about 12 times this amount, 25 barrels per capita per year, as the US consumes today. China's total annual oil consumption at 33 billion barrels would exceed that of the entire world in 2008 (31 billion barrels).

Suppose India consumed not 0.9 barrels per capita annually but about 30 times this value, placing it too on par with the rate of oil consumption in the US. With its population of 1.15 billion people, India's oil consumption would be about 29 billion barrels per year, which is a figure nearly equal to all global oil consumption in 2008. If oil consumption increases continue in developing nations, the stress placed on supplies would likely lead to a massive shortfall in global oil production.

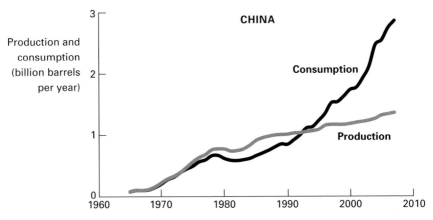

Figure 3.12   China's annual oil production and consumption. (Data: *BP Statistical Review*)

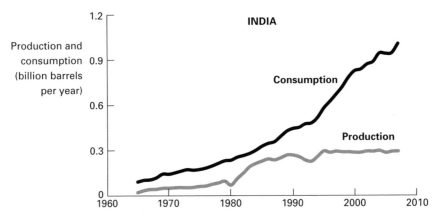

Figure 3.13   India's annual oil production and consumption. (Data: *BP Statistical Review*)

The Price of Oil is Increasing: Does This Indicate Scarcity?

Figure 3.14 shows the price of oil (not adjusted for inflation) and rate of global oil production since 1861. Over the long term, both have trended up. The price of oil doubled in 1973 in response to the OPEC oil embargo and the major oil crisis of the 1970s. The price paid for oil in dollars of the day (nominal dollars, not adjusted for inflation) dropped after the 1970s oil crisis, only to rise again after the millennium. Although the price has fluctuated, the nominal price of oil was higher in 2008 than it had ever been. But that price trend neglects inflation.

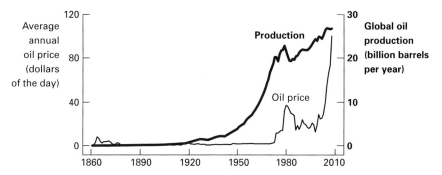

Figure 3.14   Average annual per-barrel price of oil in nominal dollars (not adjusted for inflation) and annual oil production since 1861 have both increased. Note: 2009 price decline not plotted. (Data: production, EIA; oil prices, *BP Statistical Review* and EIA)

However, considering inflation, it isn't true to say that the real price of oil has generally increased. Table 3.1 shows the purchasing power of dollars over time based on annual increases in the consumer price index (CPI). Shown is the value of $100 in 2007 during the nearly 150 years since oil has been produced.

As seen in Table 3.1, a dollar is not worth what it used to be. A purchase in 1860 of $4.00 would cost $100 in 2007. Looked at another way, if you had $100 in expenses in 1860, it would be equivalent to about $2,500 in expenses in 2007 as a result of inflation. In 1960, oil sold for about $2 per barrel, but, due to inflation, it would take about $13 in 2007 to purchase that same oil. So, although the nominal price of oil has gone up, that probably is not a good indication of much besides inflation. In part, oil costs more today simply because everything costs more.

Table 3.1   Value since 1860 of $100 in 2007[*]

Year	Equivalent $100 in 2007
1860	$4.00
1900	$4.10
1920	$9.60
1940	$6.80
1960	$14.30
1970	$18.70
1980	$39.70
1990	$63.00
2000	$83.10
2007	$100.00
2008	$103.80
2009	$103.20

[*]Inflation equivalent value of $100 in 2007 based on annual change in the US consumer price index, CPI. Inflation adjustments based on Sahr (2009).[61]

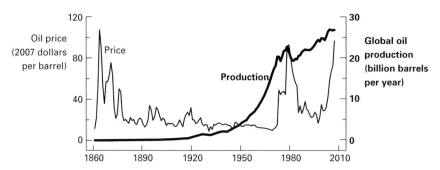

Figure 3.15   Inflation-adjusted annual average price of oil (in 2007$) and the increase in global oil production. Note that these prices are annual averages. The price of oil over any particular year has varied. For example, in 2008, daily closing oil prices ranged from $30 to $145 per barrel for West Texas Intermediate crude. 2009 price decline not plotted. (Data: oil production, EIA and Hubbert (1969)[62]; oil prices *BP Statistical Review* and EIA)

Figure 3.15 shows the price of oil adjusted for inflation based on the CPI, with values shown in 2007 dollars (2007$). Viewed this way, the post-2002 spike in prices looks as if it might really be a continuation of the trend that began in the 1974–85 period but was disrupted by the temporary weakness of OPEC when their production cutbacks were terminated. If a commodity is becoming scarce, then one would expect the inflation-adjusted price to increase. Is the increasing price trend from 2002 to 2008 a reflection of oil scarcity? The price of oil indeed climbed to an historically high level in 2008, but, given the fluctuations in the inflation-adjusted price, it is difficult to claim oil scarcity based on the high 2008 prices alone. Historical spikes in the price of oil were not because the world was running out of oil.

It is instructive to look at the history of global expenditures on oil, as seen in Figure 3.16, which shows expenditures during the most recent 30-year periods (1979–2008, 1949–78) and longer period prior to 1949. Over time and adjusted for inflation, the world has spent more and more money on oil, and this global expenditure has increased dramatically even with inflation taken into account. Since around the time oil was produced in 1861, through 2008, the world has spent $40 trillion on oil (2007$). Before 1949, the total spent on oil was "merely" $1 trillion, or less than 3 percent of the all-time total. For the three decades from 1949 to 1978, the global oil expense was $7 trillion. But these numbers pale in comparison to what has been spent from 1979 through 2008: $31 trillion, or over three-quarters of all money ever used to purchase oil globally. Remarkably, over 30 percent of all money ever spent on oil was spent during the period from 2000 through 2008: over $12 trillion (in 2007$). The US alone spent over $1 trillion dollars (2007$) on imported crude oil from 2005 through 2008. The latter figure reflects a climb in oil prices coupled with rapidly increasing global demand during this period. Whether this trend in itself is a sign of scarcity and whether such a global expense is sustainable are issues that will be revisited in the chapters ahead, in which oil expenditures will be compared to the scale of the global economy.

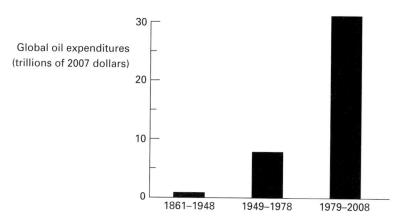

Figure 3.16 Historical global oil expenditures in inflation-adjusted dollars (2007$). (Data: *BP Statistical Review* and EIA)

Forecasts Support a Decline in Global Production Using Extensions to Hubbert's Approach

Many analysts have built on the approach introduced by Hubbert and likewise have concluded that the date of global peak oil production is nearing. Notable is the US Department of Energy (DOE) forecast method (2004). Like Hubbert's, the DOE method assumes a range of values for the global oil endowment and matches historical production using a simple mathematical expression. However, the DOE model forecasts production based on the assumption that there will be a constant annual percent growth rate in global production until a production peak is reached. After the oil peak, their model assumes that oil production will decline but that remaining reserves will always equal 10 times the global production rate (that is, there will always be a reserve-to-production ratio of 10 to 1). The net result of the DOE approach is that it predicts the date of peak oil production, but after the peak, the projected decline in oil production does not mirror the increase in production, as assumed by Hubbert. This model of declining production behavior is more physically realistic than that using a symmetric logistic (Hubbert-type) curve.

Figure 3.17 shows DOE forecasts for a range of possible values of oil production rate increases when the global oil endowment was set equal to

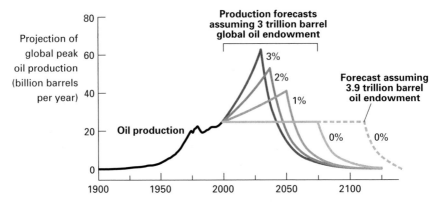

Figure 3.17   US Department of Energy projections of peak oil production under different production increase scenarios. Shown are projections based on production growth rates of 0 to 3 percent, assuming the best estimate of the oil endowment, 3 trillion barrels, by the USGS in their 2000 Assessment. The forecast shown as a dashed line is based on no production growth (0 percent) and assumes the highest estimate of the oil endowment by the USGS of 3.9 trillion barrels (after Wood et al., EIA (2004)[65])

the 3 trillion barrels estimated by the USGS in its 2000 Assessment. Although not all of the projections are shown in Figure 3.17, the DOE forecasts did consider a range of oil endowment values from 2,248 to 3,896 billion barrels based on the uncertainty estimates given in the USGS Assessment. Oil production rate increases were varied from 0 to 3 percent based on the DOE's judgment of likely values. For scenarios of 0 to 3 percent production growth, the DOE's results suggest that peak oil might occur anywhere from 2030 to 2075 assuming the best estimate of the oil endowment value given by the USGS. In the most optimistic case of no growth in production and assuming the highest estimate of the oil endowment, a permanent decline in production would begin in 2112. Most of the DOE forecasts in their study support the concern that within about two generations, global peak oil production will be reached and will be followed by rapidly declining production.[63,64]

Summary

Beyond the simple analysis that Hubbert presented some 50 years ago, the case for global oil depletion appears to be compelling when viewed from a variety of perspectives involving both the sources of oil supply and global demand.

- Oil production has declined in other countries besides the US.
- Oil production has exceeded new oil discoveries.
- US peak oil production lagged peak oil discoveries by about 35 years. The global oil discovery peak occurred in the early 1960s.
- Fewer global giant oil fields are being discovered and about 80 percent of those developed before 2005 show production declining at an annual rate of about 6 percent.
- Developing nations, such as China and India, have steadily increased their use of oil and import significant quantities. Projecting their future demand suggests great strain on future global oil supplies.
- Reported reserve and oil endowment values may be inflated to begin with.
- Global oil expenditures during the past 30 years are four times those in the prior 30-year period.
- Under assumptions that many would consider reasonable, the US DOE projects that peak oil production might occur by 2030 under their high growth scenario, with rapid declines in production thereafter.

Notes and References

1. Malthus, T. R. (1798). *An Essay on the Principle of Population*, 1993 imprint edited by G. Gilbert. Oxford University Press, Chapter 1: 13–14.
2. Ibid., Chapter 7: 61.
3. Sen, A. (1982). "The food problem: Theory and Policy," *Third World Quarterly*, **4**(3): 447–59.
4. Based on US Census Bureau (2008). Historical Estimates of World Population, citing United Nations, 1999, *The World at Six Billion*, Table 1, "World Population From Year 0 to Stabilization," page 5, www.un.org/esa/population/publications/sixbillion/sixbilpart1.pdf, and 2009 www.census.gov/main/www/popclock.html. In his *An Essay on the Principle of Population*, published in 1798 (1993 Oxford University Press imprint edited by G. Gilbert (ed.), Chapter 7), Malthus presented a similar example of the impact of exponential population growth to that Hubbert used some 150 years later. Malthus states: "This ratio of increase, though short of the utmost power of population, yet as the result of actual experience, we will take as our rule, and say that population, when unchecked, goes on doubling itself every twenty-five years or increases in a geometrical ratio. ... Let us now take any spot of earth, this Island [the UK] for instance, and see in what ratio the subsistence it affords can be supposed to increase. We will begin with it under its present state of cultivation. ... The population of the Island is computed to be about seven millions, and we will suppose the present produce equal to the support of such a number. ... And at the conclusion of the first century the population would be one hundred and twelve millions and the means of subsistence only equal to the support of thirty-five millions, which would leave a population of seventy-seven millions totally unprovided for."
5. Turner II, B. L., R. E. Kasperson, P. A. Matson, J. J. McCarthy, R. W. Corell, L. Christensen, N. Eckley, J. X. Kasperson, A. Luers, M. L. Martellog, C. Polsky, A. Pulsiphera, and A. Schiller (2003). "A framework for vulnerability analysis in sustainability science," *Proceedings of the National Academy of Science*, July 8, 2003, **100**(14): 8074–9; United Nations Food Programme (2008) www.wfp.org/aboutwfp/introduction/hunger_stop.asp?section=1&sub_section=1
6. Meadows, D., D. Meadows, J. Randers, and W. W. Behrens (1972). *The Limits to Growth*. Potomac Associates Books: 205.
7. Ibid.
8. Ivanhoe, L. F. (1997). *Hubbert Center Newsletter*, #97/1.
9. Laherrère, J. H. (2001). "Energy resources – cornucopia or empty barrel?: Discussion," *AAPG Bulletin*, **85**(6) June 2001: 1083–7.
10. Tsoskounogloua, M., G. Ayerides, and E. Tritopoulou (2008). "The end of cheap oil: Current status and prospects," *Energy Policy*, **36**(10): 3797–806.

11. McCabe, P. J. (2001). "Energy resources – cornucopia or empty barrel?: Reply," *AAPG Bulletin*, **85**(6) June 2001: 1093–7, citing Campbell, C. J. (1997), *The Coming Oil Crisis*. Brentwood, UK: Multi-Science Publishing Company and Petroconsultants, 210 pp.

12. Hirsch, R. L., R. Bezdek, and R. Wendling (2005). *Peaking of World Oil Production: Impacts, Mitigation, and Risk Management*, Department of Energy National Energy Technology Laboratory, February 2005.

13. Fanning, L. M. (1950). "A Case History of Oil-shortage Scares," in *Our Oil Resources*, second edition, edited by L. M. Fanning. New York: McGraw-Hill, 306–406.

14. Olien, D. D., and R. M. Olien (1993). "Running Out of Oil: Discourse and Public Policy, 1909–1929," *Department of Business and Economic History*, **22**(2) winter 1993, citing Response by Secretary of Interior to Senate Resolution, which appears in US Senate, Document 310, 64th Congress, First Session, February 2, 1916.

15. Fanning, L. M. (1950). "A Case History of Oil-shortage Scares," in *Our Oil Resources*, second edition, edited by L. M. Fanning. New York: McGraw-Hill, 314.

16. US Air Museum, 2007, Duxford, UK.

17. Fanning, L. M. (1950). "A Case History of Oil-shortage Scares," in *Our Oil Resources*, second edition, edited by L. M. Fanning. New York: McGraw-Hill, 327–8.

18. Porter, E. D. (1995). American Petroleum Institute, "Are We Running Out of Oil?" American Petroleum Institute, Discussion Paper #081.

19. Ibid.

20. Ibid.

21. Based on Yergin, D. (1991). *The Prize: The Epic Quest for Oil, Money, and Power*. New York: Free Press, Simon and Schuster: 588–698.

22. Energy Information Administration, Table 11.7 Crude Oil Prices by Selected Type, 1970–2006, www.eia.doe.gov/emeu/aer/txt/ptb1107.html

23. Yergin, D. (1991). *The Prize: The Epic Quest for Oil, Money, and Power*. New York: Free Press, Simon and Schuster.

24. Watkins, G. C. (2006). "Oil scarcity: What have the past three decades revealed?" *Energy Policy*, **34**: 508–14, citing F. Parra, 2004, *Oil Politics – A Modern History of Petroleum*, I. B. Tauris.

25. Porter, E. D. (1995). American Petroleum Institute, "Are We Running Out of Oil?" American Petroleum Institute, Discussion Paper #081.

26. Adelman, M. A. (1995). *The Genie Out of the Bottle: World Oil Since 1970*. Cambridge, MA, MIT Press: 178, citing CIA, "The Geopolitics of Energy," statement submitted to the Committee on Energy, US Senate, reprinted in *Petroleum Intelligence Weekly*, May 19, 1980: S1–S4.

27. The CPI climbed from 39.8 to 86.3 between 1971 and 1980, ftp://ftp.bls.gov/pub/special.requests/cpi/cpiai.txt

28. Energy Information Administration (2001). *Future Production for the North Slope of Alaska*, 74 pp.
29. Government Accountability Office (2007, February). *CRUDE OIL Uncertainty about Future Oil Supply Makes It Important to Develop a Strategy for Addressing a Peak and Decline in Oil Production*; Hirsch, R. L., R. Bezdek, and R. Wendling (2005). *Peaking of World Oil Production: Impacts, Mitigation, and Risk Management*, Department of Energy, National Energy Technology Laboratory, February 2005; Kerr, R. A. (2007). "The Looming Oil Crisis Could Arrive Uncomfortably Soon," *Science*, **316**, April 20, 2007: 351.
30. BP data for 2007 from *BP Statistical Review of World Energy*, June 2008; EIA (2008). *World Energy Outlook 2008*.
31. Sandrea, R. (2006). "Early New Field Production Estimation Could Assist in Quantifying Supply Trends," *Oil and Gas Journal*, May 22, 2006.
32. EIA data.
33. Discovery data from EnergyFiles Ltd. Courtesy of Michael Smith, Chief Executive, (2008). These data were smoothed using a five-year running mean. The values are estimates based on many varied and some questionable data sets. Although the trends are probably accurate, actual discovery (produced) volumes will not be fully known until all fields are abandoned far in the future. They do not include unconventional oils, such as oil sands in Canada and oil shales.
34. Campbell. C. J. and J. H. Laherrère (1998). "The End of Cheap Oil," *Scientific American*, March 1998: 78–83.
35. Iran again increased its reserves by 40 percent in 2004 (EIA data, 2008).
36. Laherrère (2006) presents a more sophisticated analysis that considered a multi-peak Hubbert model with a similar projection of peak oil production and decline. Laherrère, J. H. (2006). "Learn strengths, weaknesses to understand Hubbert curve," *Oil and Gas Journal*, August 2, 2006.
37. Hubbert, M. K. (1969). "Energy Resources," in *Resources and Man*, W. H. Freeman and Co., Chapter 8: 157–242.
38. United States District Court for the Southern District of Texas, Houston Division, Securities and Exchange Commission, plaintiff, *v.* complaint, Royal Dutch Petroleum Company and: H-04-3359 the "Shell" Transport and Trading Company, plc, Defendants. United States of America before the Securities and Exchange Commission, Securities Exchange Act of 1934 Release No. 50233 / August 24, 2004, Accounting and Auditing Enforcement Release No. 2085 / August 24, 2004, Administrative Proceeding File No. 3-11595, In the Matter of Royal Dutch Petroleum Company and The "Shell" Transport and Trading Co., plc. Respondents Order Instituting Cease-and-Desist Proceedings Pursuant Section 21C of the Securities Exchange Act of 1934, Making Findings, and Imposing a Cease-and-Desist Order; BBC News: "Shell escapes charges on reserves," June 30, 2005; "How much oil do we really have?" 15 July, 2005; "Shell settles oil reserve claims," November 4, 2007.
39. Horn, M. K. (2006a). "Giant fields 1868–2003," Data on CD-ROM, in M. Halbouty (ed.), "Giant oil and gas fields of the decade 1990–1999," AAPG

Memoir 78, 2003, 340 pp., modified November 2006 to reflect giant oil discoveries 2000 to 2006. Further updates, personal communication, 2008 and 2009.

40. Horn, M. K. (2007). "Giant fields likely to supply 40%+ of world's oil and gas," *Oil and Gas Journal*, April 9, 2007: 35–7; Horn, M. K. (2009). www.sourcetoreservoir.com

41. IFP Panorama Technical Reports (2005). *A Look at New Oil and Gas Discoveries*, www.ifp.com/information-publications/notes-de-synthese-panorama/panorama-2005

42. Bahorich, M. (2006). "End of oil? No, it's a new day dawning," *Oil and Gas Journal*, **104**(31), August 21, 2006: 30–4.

43. Horn, M. K. (2006b). "World giant oil discoveries seem not to be at an end," *Oil and Gas Journal*, November 6, 2006: 33.

44. Data through 2008 from Horn (2009), www.sourcetoreservoir.com

45. Horn, M. K. (2009). www.sourcetoreservoir.com

46. Ibid.

47. Robelius, F. (2005). "Giant Oil Fields of the World," Presentation AIM Industrial Contact Day, May 23, 2005.

48. www.glossary.oilfield.slb.com/Display.cfm?Term=water%20cut

49. Nasser, A. H., and N. G. Saleri (2004). "Reserves and Sustainable Oil Supplies: Role of Technology and Management," 5th International Oil Summit, Paris, France, April 29, 2004.

50. Simmons, M. R. (2005). "Twilight in the Desert," *World Energy*, **8**(2): 44–51.

51. Höök, M., R. Hirsch, and K. Aleklett (2009). "Giant oil field decline rates and their influence on world oil production," *Energy Policy*, **37**(6): 2262–72.

52. Robelius, F. (2007). "Giant Oil Fields – The Highway to Oil: Giant Oil Fields and their Importance for Future Oil Production." Doctoral thesis, Uppsala University, Sweden.

53. International Energy Agency (2008). *World Energy Outlook 2008*.

54. Höök, M., R. Hirsch, and K. Aleklett (2009). "Giant oil field decline rates and their influence on world oil production," *Energy Policy*, **37**(6): 2262–72; IEA (2008) *World Energy Outlook 2008*.

55. Discovery data from Klett (2003). Klett, T. R. (2003). "Graphical Comparison of Reserve-Growth Models for Conventional Oil and Gas Accumulations," Chapter F of Geologic, Engineering, and Assessment Studies of Reserve Growth, edited by T. S. Dyman, J. W. Schmoker, and M. Verma, U.S. Geological Survey Bulletin 2172–F.

56. Laherrère, J. H. (2000). "Learn strengths, weaknesses to understand Hubbert curve," *Oil and Gas Journal*, **98**(16), April 17, 2000.

57. Klett, T. R. (2003). "Graphical Comparison of Reserve-Growth Models for Conventional Oil and Gas Accumulations," Chapter F of Geologic, Engineering, and Assessment Studies of Reserve Growth, edited by T. S. Dyman, J. W. Schmoker, and M. Verma, U.S. Geological Survey Bulletin 2172–F.

58. Bahorich, M. (2006). "End of oil? No, it's a new day dawning," *Oil and Gas Journal*, **104**(31), August 21, 2006: 30–4.

59. Discovery data from EnergyFiles, Ltd. Courtesy of Dr. Michael Smith, Chief Executive, 2008. These data were smoothed using a five-year running mean. The values are estimates based on many varied and some questionable data sets. Although the trends are probably accurate, actual discovery (produced) volumes will not be fully known until all fields are abandoned far in the future. They do not include unconventional oils, such as oil sands in Canada and oil shales. Discovery data shown are similar to those in P. R. A. Wells (2005). "Oil supply challenges – 1: The non-OPEC, oil and gas decline," *Oil and Gas Journal*, **103**(7), February 21, 2005 (after Harper (2003), *Petroleum Review*).

60. *World Fact Book*, 2008, www.cia.gov/library/publications/the-world-factbook/geos/ch.html; and United Nations, Division of Economic and Social Affairs, 2008, http://esa.un.org/unpp/index.asp

61. Sahr, R. (2009). "Inflation conversion factors for dollars, 1774 to estimated 2018," http://oregonstate.edu/cla/polisci/faculty-research/sahr/sahr.htm

62. Hubbert, M. K. (1969). "Energy Resources," in *Resources and Man*, W. H. Freeman and Co., Chapter 8: 157–242.

63. Wood, J. H., G. R. Long, and D. F. Morehouse, (2004). *Long-Term World Oil Supply Scenarios: The Future Is Neither as Bleak or Rosy as Some Assert.* Energy Information Administration. Feature article.

64. Mohr, S. H. and G. M. Evans (2007). "Mathematical model forecasts year conventional oil will peak," *Oil and Gas Journal*, **105**(17), May 7, 2007. Mohr and Evans (2007) present a model based on equations that are more general than those used by Hubbert. Their model distinguishes among production by OPEC, the former Soviet Union, and the rest of the world. They estimate that global peak oil will occur in 2012 or 2024 for two different scenarios in which the oil endowment is assumed to be 2.2 or 2.7 trillion barrels. Their model allows for asymmetry in pre-peak and post-peak production.

65. Wood, J. H., G. R. Long, and D. F. Morehouse, (2004). *Long-Term World Oil Supply Scenarios: The Future Is Neither as Bleak or Rosy as Some Assert.* Energy Information Administration. Feature article.

4

Counter-Arguments to Imminent Global Oil Depletion

The arguments pointing to the imminent depletion of oil resources are seemingly convincing. Many have been accepted as common wisdom. However, there is another side to the oil-depletion story. The counter-arguments are pointed critiques of the assumptions and methods of Hubbert and those who have followed in his footsteps, along with their analyses, assumptions, and interpretations of data.

Myth I: Hubbert's Predicted Production Rates Were Accurate

Much has been made of Hubbert's approach based largely on the accuracy of his prediction of oil production decline in the coterminous US and his further prediction of global "peak oil" and production decline. Now that over five decades have passed since Hubbert's first predictions, let us see how well he did. Any useful measure of his approach must be based on how accurately production rates, unknown at the time, were predicted. One should not be subliminally impressed with a match to historical production data, because the Hubbert curve is forced to fit those data. The fit is forced primarily by adjusting the parameter controlling the slope of the "early time" rising limb of the curve, with the constraint imposed by the total oil endowment, which is the area under the curve.

Oil Panic and the Global Crisis: Predictions and Myths. 1st edition. By Steven M. Gorelick.
Published 2010 by Blackwell Publishing, ISBN 978-1-4051-9548-5 (hb)

US Oil Production

The headline of the March 16, 1956, issue of *Petroleum Week* read, "Is Oil Nearing a Production Crisis?" and stated in bold lettering, "A prominent Texas geologist predicted last week that the upward spiral of U.S. production may end about 1965, to be followed by a decrease of from 5% to 10% a year thereafter." The prominent Texas geologist was M. King Hubbert (Figure 4.1). Since that time, Hubbert's curve and the success of his prediction of US peak oil production is found in numerous books and articles and is taught in many college courses covering the topic of Earth resources. The followers of Hubbert are many and are adamant that his method is sound and accurate. They even have their own society: The Association for the Study of Peak Oil.

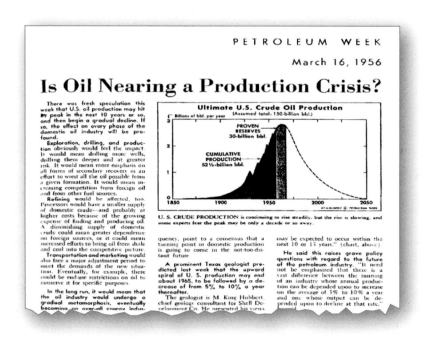

Figure 4.1   Article from *Petroleum Week* in 1956 describing Hubbert's prediction of the decline in US oil production and showing peak production in 1965, assuming an oil endowment of 150 billion barrels.

Over 50 years have passed since this first of Hubbert's predictions, and we can evaluate it against actual US oil production data. The graph that is often shown by followers of Hubbert is reproduced below (Figure 4.2). This figure shows actual oil production in the coterminous US superimposed on Hubbert's 1956 predictions, which included a best prediction, based on a 150 billion barrel US oil endowment, and a high prediction, based on a 200 billion barrel endowment. The latter prediction turned out to be the most accurate. Hubbert predicted a peak in oil production and the general pattern of decline. His logistic curve seems to capture the essence of the production trend when applied to all of the production data, including the post-peak data. In many ways, this was a remarkable feat. However, there is reason to remain skeptical about inferring that Hubbert's success in the US means that predicting global oil production trends will be similarly successful.

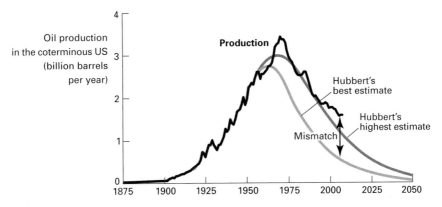

Figure 4.2 Hubbert's 1956 predictions of coterminous US oil production. Hubbert made what he termed his best (most likely) prediction assuming 150 billion barrels for the US oil endowment and a high prediction assuming a 200 billion barrel oil endowment. His best prediction underestimated peak oil production. Only his high estimate came close and predicted the timing of the peak (after Deming, 2003[1]). (Data: production from EIA and curves from Hubbert (1956)[2])

Much has been made of the accuracy of Hubbert's predictions of US oil production, but on closer inspection, what can be learned? David Deming, a professor at the University of Oklahoma, dug into the numbers, conducted a post-audit, and showed that Hubbert's predictions were not so amazing after all.[3,4] Based on data available in 1956, Hubbert adopted the value of 150 billion barrels as his best estimate of the US oil endowment. As shown in Figure 4.2, Hubbert's prediction for the time of peak oil in the coterminous US was not very accurate using his best estimate of the oil endowment. Hubbert posited that the time of "peak oil" production would be 1965, only nine years after his published predictions in 1956. Instead, the peak occurred in 1970, some 14 years later, and Hubbert had under-predicted the peak value by over 25 percent. Inspecting more recent data available since Deming's analysis (including data beyond 2000), one can see an inaccuracy in Hubbert's early prediction that is perhaps more glaring than his estimates of time to peak oil and value of peak oil: Hubbert's projected oil production rate in 2008 was 0.5 billion barrels per year versus the true value that we now know to be about 1.55 billion barrels per year. Hubbert was off by a factor of three. According to Hubbert's prediction based on his favored endowment value, oil production in 2009 in the coterminous US should be at the level it was in 1922. That is not the case.

In addition to what he considered to be his best estimate of the US lower 48 states' oil endowment, Hubbert covered his bets by adopting an upper bound on the value, which was little more than a wild guess on his part, and redid his estimate. As Deming points out, Hubbert added "an amount equal to 8 east Texas oil fields," or 50 billion barrels, to his estimate of the US oil endowment. As noted in a 1981 *Science* article by Hall and Cleveland,[5] Hubbert backed up his estimate of US ultimate oil by citing a significant decline in oil discovered per foot of exploratory drilling. In the 1930s, the value was 250 barrels per foot, but by the 1950s, this value had dropped to merely 40 barrels per foot. Based on this drilling-to-discovery ratio, Hubbert extrapolated from data to predict that the lower 48 states' oil endowment would be between 150 and 200 billion barrels. When one checks Hubbert's approach based on the drilling-to-production ratio, the method fails because that trend did not persist. The ratio increased in the 1960s but then declined precipitously in the 1970s to as low as 15 barrels per foot.

With his inflated estimate of 200 billion barrels (versus 150 billion), Hubbert's 1956 prediction better resembles the actual amount and time of peak oil production. Yet even using the figure of 200 billion barrels, when projected to the beginning of 2009, Hubbert's production estimate falls short. As Deming noted, it is not clear that Hubbert knew which prediction was better even after the actual occurrence in 1970 of "peak oil." In 1980, Hubbert made yet another (final and worse) prediction, this time based on an oil endowment of 170 billion barrels. This last forecast under-predicted actual production at the beginning of 2009 by a factor of 2.5 (Figure 4.3).

Hubbert did not use the logistic (bell-shaped) curve to make his predictions in 1956.[8] Rather, he drew the curves by hand. The hand-drawn nature of the curves is evident, as his curves were not symmetric, and some showed a long trailing tail after "peak oil." Hubbert explained:

> In my figure of 1956, showing two complete cycles for U.S. crude-oil produc-
> tion, these curves were not derived from any mathematical equation. They were
> simply tailored by hand subject to the constraints of a negative-exponential

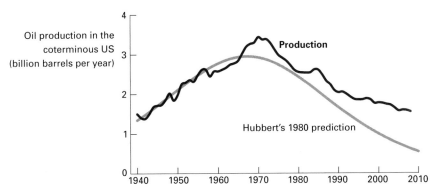

Figure 4.3 Hubbert's 1980 prediction using a US oil endowment value of 170 billion barrels. The actual production value at the beginning of 2009 was 2.5 times that of Hubbert (after Deming (2000)[6]). (Data: production from EIA and curve from Hubbert (1982)[7])

decline and a subtended area defined by the prior estimates for the ultimate production. Subject to these constraints, with the same data, I suggest that anyone interested should draw the curves himself.[9]

There is a great degree of arbitrariness of a hand-drawn curve, particularly when one does not know the trajectory of post-peak production decline, and the value of the area under the curve, the oil endowment, is so uncertain. Some of the arbitrariness was later eliminated when Hubbert adopted the use of the logistic curve rather than drawing curves by hand, but the problem of determining the value of the oil endowment still remained.

Retrospectively, it is a simple matter to apply Hubbert's logistic curve to the US production data through the present. Figure 4.4 shows the curve often touted by the followers of Hubbert. It is a fit to all of the data, both before and after the peak. Matching the peak in retrospect is a lot easier than predicting the peak before it occurs. Such "post-peak" predictions differ from Hubbert's forecast in that Hubbert never predicted a peak value as high as the actual peak. With this retrospective fit, Hubbert's approach is easily able to match much of the historical production data, including the peak. Yet even with the advantage of post-peak hindsight, the "nowcast" does not match actual production data for the past 10 years and substantially underestimates production, as displayed in Figure 4.4 (for 2008, applying Hubbert's method gives a predicted value of 0.92 billion barrels per year versus the actual 1.55 billion barrels per year). This figure presents an enlargement of the nowcast prediction for the years 1994 through 2008, showing the descending limb of the predicted curve versus actual production data. For this period, the prediction shows a declining trend that is three times the actual downward slope based on oil production data. The Hubbert approach using modern production data not available to Hubbert in 1956, but with his highest estimate of the coterminous US oil endowment, forecasts that over 90 percent of the US endowment has been depleted (2008 prediction of total oil produced is 183 of 200 billion barrels, leaving less than 20 billion barrels). The US has maintained reserves of over 20 billion barrels since records were kept, and reserves do not include undiscovered oil in new or existing areas or oil that becomes profitable to produce if the price goes up. The approach also predicts 95 percent US oil depletion by 2018. By this accounting, production in the US lower 48 states falls to less than a third of its current level after 2018. The actual production data do not support such a trend.

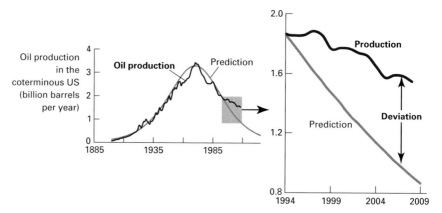

Figure 4.4   Curve fit to coterminous US oil production data through 2008 using Hubbert's approach to pre- and post-peak data. The figure shows the deviation of the prediction from actual oil production during the past decade. (Data: EIA)

Based on the Department of Energy's EIA oil production data, 183 billion barrels were produced through 2008 in the coterminous US, with another 14 billion barrels produced in Alaska. Yet the latest USGS endowment figure of 330 billion barrels suggests that in fact almost half of the US endowment remained in 2008. Other estimates of the US oil endowment are even higher than that of the USGS. William L. Fisher of the University of Texas, Bureau of Economic Geology, has placed ultimate recovery at 368 billion barrels.[10] In March 2006, the US DOE maintained that there remains 430 billion barrels of technically recoverable oil in the US.[11] The vast majority of that lies in the lower 48 states. Either the USGS, Fisher, and DOE values are utterly wrong, or US production can continue for a long time to come. If the coterminous US average production remained constant at the 2008 value of 1.55 billion barrels per year, which is half that in Hubbert's most accurate peak prediction, production could continue for almost 100 years based on the USGS endowment figures. This projection does not rely on the higher available estimates or further upward revision of the oil endowment based on new technologies for oil recovery or existing oil resources made into profitable reserves by rising oil prices, as discussed later.

The Bell-Shaped Curve

An appeal of Hubbert's original approach is that it is based on a mathematical model that shows a familiar bell-shaped curve. Even with its predictive failure over the past decade, the claim of success of Hubbert's approach has been US oil production based on this curve. The bell curve is familiar as it is often used to represent the frequency of a quantity (statistical population): for example, the distribution of heights of a large number of women in a theater. However, why should the global oil production trajectory correspond to a bell-shaped distribution? There is no particularly compelling reason for the rise to mirror the decline in production. Oil production curves for individual well fields do tend to show rapid production that peaks and then tails off with time. However, even when production curves show a decline, they do not generally exhibit the nice symmetry of a logistic curve. For entire nations, continents, or the world, a symmetric bell-shaped production curve is not expected. On a global basis, only 8 of 51 non-OPEC nations' production curves follow a bell curve.[12] As noted by USGS scientist Ronald Charpentier:

> Symmetry of the production curve is a very strong assumption. The assumption that addition of production curves leads to a symmetrical bell-shaped curve requires that the curves be added randomly in time. Production history curves are not added randomly in time, but rather new plays must be added to the future part of the curve, skewing the curve.[13]

Although Hubbert and some of his followers distinguish between their use of a logistic curve and use of the bell-shaped curve from statistics for a normal distribution,[14] a former Princeton geology professor, Kenneth Deffeyes, uses a normal distribution to represent oil production, with no theoretical justification. Professor Deffeyes, who predicted that a decline in global oil production would be under way by 2006, is an outspoken proponent of Hubbert's approach.[15] The underlying idea of using a normal, or Gaussian, distribution is that the sum of a bunch of random distributions can result in a normal distribution according to a statistical principle called the central limit theorem (CLT).

> The CLT acts to create "bell-shaped" distributions when distributions that are independent of one another are summed. While individual production curves are summed, they are not independent. Production at a given oil field is determined at least in part by the decisions of the producers. These producers, across regions, nations, and even at a global level, respond to common stimuli. At a regional level, common stimuli include local transport costs, availability of nearby markets, and regulatory pressures (such as state or provincial environmental mandates), while national politics can force production up or down, particularly in nations with central control over production (e.g., OPEC). And, of course, both long- and short-term trends influence producers simultaneously across the globe. Thus, there is no theoretical reason to expect Gaussian production profiles to fit all cases.[16]

It might be thought that due to averaging effects, an aggregate of smaller regions into a larger region is more likely to exhibit a bell-shaped production history than smaller regions. However, one study of records from 139 oil-producing regions found that, "… regions of larger area do not adhere to the Hubbert model more strongly than smaller regions."[17] As shown in Figure 4.5, North America has a fairly stable rate of oil production. Europe is the only large region where oil production values show significant modern decline, and it provides less than 7 percent of world supply.

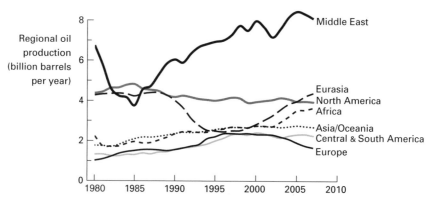

Figure 4.5   Regional oil production since 1980. Modern production shows a declining trend in Europe but not in other regions. (Data: EIA)

US Natural Gas Production

Hubbert also applied his method to US natural gas production data available in 1956. As show in Figure 4.6, which updates Deming's comparison of 2000, Hubbert's prediction significantly underestimates levels of production since the 1970s. Production levels in 2008 were three times those predicted by Hubbert, and according to the Hubbert-curve prediction, the US should be producing at about the meager rates of the early 1950s. Rather, there has been a moderate upward trend in production since the mid-1980s, without the sustained decline from 1970 on that Hubbert predicted.

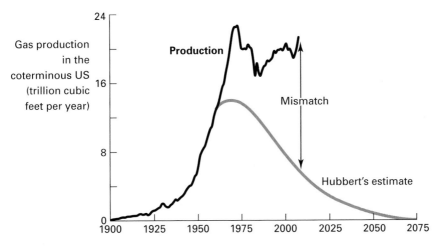

Figure 4.6 Comparison of 1956 Hubbert's prediction of US natural gas production with actual natural gas production data showing a significant mismatch. Hubbert's original curve in 1956 was drawn by hand (after Deming (2000)[18]). (Data: production from EIA and curve from Hubbert (1956)[19])

Global Oil Production

In 1956, Hubbert ventured to apply his approach to the production of all oil on Earth.[20] He assumed an oil endowment value (the sum of historical cumulative production, known reserves, and projected discoveries) of 1.25 trillion barrels and predicted that the time of "peak oil" would occur near the millennium at a peak production value of 12.5 billion barrels per year (Figure 4.7). Comparing actual production data with his predictions, we find that no peak has yet occurred, production is about 27 billion barrels per year (more

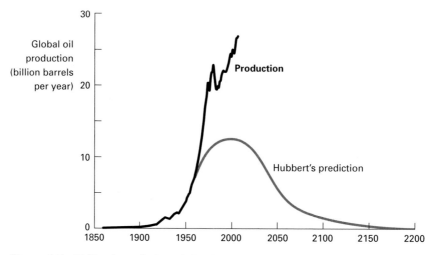

Figure 4.7   Hubbert's prediction of global oil production made in 1956 drastically underestimated modern production of about 27 billion barrels per year, with no peak in production yet occurring. Hubbert assumed a global oil endowment of 1.25 trillion barrels, which is far less than the current USGS estimate of over 3 trillion barrels. (Data: production from EIA and curve from Hubbert (1956)[21])

than double his prediction), and the current USGS estimate of the oil endowment is over 3 trillion barrels (almost 2.5 times that of Hubbert).

As a test of the validity of Hubbert's predictive method, even had he used more recent estimates of the global oil endowment, it is a simple matter to apply his approach to the historical production data through 1980 and then look at how well it predicts the next 28 years of production. Forecasts are shown in Figure 4.8 under assumed global oil endowment values of 1.25 trillion barrels (used by Hubbert in his 1956 prediction), 2.1 trillion barrels (as he assumed in a 1971 prediction), and 3 trillion barrels (the latest USGS estimate of the global oil endowment). None of the forecasts made with Hubbert's approach compares favorably with the production data after 1980. These data indicate no smooth trajectory toward an obvious peak of oil production or a decline as suggested by all of the curves based on Hubbert's approach. Global predictions based on Hubbert's approach fail.[22]

As shown above for global oil production, an accurate forecast using Hubbert's method should not rely on the US experience. The US is unique in that it has been explored for oil more heavily than any place on Earth. In fact, "… in Texas alone nearly 1 million wells have been drilled, against 2,300 in Iraq"[23] (Iraq is 0.7 times the size of Texas but Texas has 435 times the number of wells). The US, with 1.6 percent[24] of worldwide reserves in 2008, had

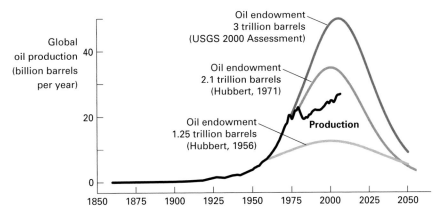

Figure 4.8 Applying Hubbert's method to production data through 1980, it is apparent that predicted values deviate significantly from the post-1980 production data, even when estimates of the global oil endowment are much greater than Hubbert initially surmised in 1956. (Data: EIA and endowment values from sources shown)

35,000 oil fields, while in the rest of the entire world, which claims over 98 percent of global reserves, only 12,500 oil fields had been developed.[25] As of the end of 2008, there were close to 500,000 producing oil wells in the US versus about 870,000 in the entire world. In all OPEC nations combined, there were about 38,000 producing oil wells (only one-fourteenth of the number in the US). OPEC member Saudi Arabia, which has the largest oil reserves in the world (20 percent), has only 1,560 producing wells.[26] It is an understatement to note that the US is a well-developed oil region. Yet the US is still highly productive and is the third largest oil producer in the world. Predicting global peak oil based on the US experience and a simple logistic-curve-fitting approach has not worked so far. Predictions claiming global production peaks have come and gone.

The above discussion is not intended to discredit Hubbert's scientific expertise. He put his assumptions, model, and predictions on the table so that they could be analyzed by others. That is what scientists are supposed to do. However, there are two important observations that one can make based on Deming's 2000 and 2003 post-audits and the follow-up with additional data shown here. First, the followers of Hubbert have tended to exaggerate the virtues and accuracy of his approach by selectively adopting certain fits to the data and estimates of the oil endowment to prove their point. This is bad science. Second, predictions of global oil and natural gas production using Hubbert's method have consistently under-predicted actual production and falsely forecast a production peak, at least when such forecasts are compared with what since have become historical data. Analysts using Hubbert's approach have the habit of ignoring their earlier predictions of the time of peak oil every so often and providing later and later dates based on assuming larger and larger values of the global oil endowment. Indeed, Hubbert had difficulty estimating the global endowment, and his various values were all much lower than today's best estimates. The consistent pessimism of the neo-Malthusian oil analysts, which results in their premature predictions of the end of economically and technically recoverable oil, is a theme that will be revisited later. Fortunately, Hubbert did not seem to take himself as seriously as his disciples have. This is suggested by Hubbert's comment in 1959: "The art of soothsaying, although probably not the world's oldest profession, can certainly offer strong claims for being the second oldest."[27]

Myth II: A Decline in Production Necessarily Indicates Scarcity

One implicit assumption of the prediction of catastrophic oil-resource depletion is that oil supply guides demand. That is, we rapidly use whatever oil we produce, which suggests that, as production rates diminish after "peak oil," there will not be enough oil to meet rising world demand. Let us look at resource scarcity as it relates to production over time.

Commodity Scarcity

Does declining production really signal resource depletion? What can be learned from the production rates of other commodities? Consider a renewable commodity, such as an agricultural product like tobacco. Figure 4.9

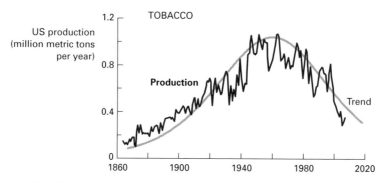

Figure 4.9   US tobacco production data and trend curve. (Data: US Department of Agriculture)

shows the US production of tobacco and a fit of a logistic curve (the type that Hubbert used in his analysis of oil production) to the annual production data. "Peak tobacco" production occurred in the 1960s (about 1962), and by 2000 production fell to a level not seen since the mid-1930s. Nevertheless, does anyone believe that the world is running out of tobacco? The flip side of production is consumption. Figure 4.10 shows the consumption of tobacco in the US. The pattern of consumption follows that of production, with exports from the US making up the difference. In the case of tobacco, it is not difficult to make the link between declining production and diminishing consumption. Production fell because peak levels could not be sustained by the post-peak demand. The trend of decreased demand among more health-conscious consumers meant lower production was warranted.

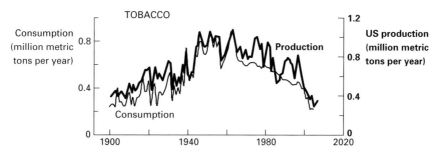

Figure 4.10   The pattern of US tobacco production follows US tobacco consumption. (Data: US Department of Agriculture)

Production of a given commodity might follow the trend of a logistic curve, as Hubbert ultimately relied on for his final predictions, but there is no particular reason why the rise and fall in production should be symmetric (mirror) images, which is a property of the logistic bell-shaped curve. Consider flaxseed as another example. As shown in Figure 4.11, although a logistic curve would reproduce the overall shape of the production history, the decline in US production was relatively protracted compared with the rate of increase that occurred before the peak in 1948. There was also a precipitous decline in post-1970 production to pre-1935 levels. Whatever the declining trend, there is no suggestion that there is a shortage of flaxseed in the US. The point is that a declining production curve obviously does not equate to resource depletion in these cases.

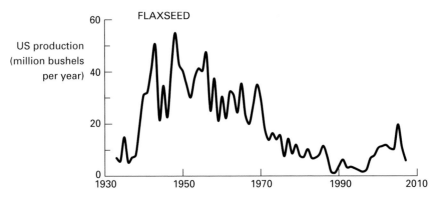

Figure 4.11   US production of flaxseed. (Data: US Department of Agriculture)

Tobacco and flaxseed are examples of renewable commodities, ones that can be grown and replenished. It is easy and perhaps even logical to dismiss these examples as inapplicable to the analysis of a non-renewable resource like oil. However, each of these examples demonstrates that demand for a commodity, rather than depletion, has resulted in declining production. A better question is this: Does a decline in production of non-renewable commodities, like metals, follow a different pattern?

Consider zinc, which is a relatively abundant element in Earth's crust. Zinc has been actively mined in the US since the 1900s (see Figure 4.12). Early demand was highest in the agricultural and chemical industries (including paint- and rubber-making), but in modern times it is used largely by steel companies, mainly for galvanizing their products and creating alloys with other metals. The US has 20 percent of the world's reserve base,[28] or 90 of 460 million metric tons. This US reserve base figure is about 100 times US peak zinc annual production and about 500 times current annual production from US mines.

The US is a major user of zinc. However, even though the US has large zinc reserves, it produces less than a third of the zinc that it consumes. What accounts for the decline in US production? The answer lies in worldwide production data. In the 1940s as the US approached "peak zinc" production, worldwide production took off (Figure 4.13) and it became cheaper to import zinc than to mine it at home. In the case of zinc, declining US production was driven by cost, not commodity depletion. Given the world's zinc resources of 1.9 billion metric tons, it is more economical in the US to import most of the processed zinc metal, primarily from Canada and Mexico, and to obtain 90 percent of imported raw ore from Peru, Australia, and Ireland.

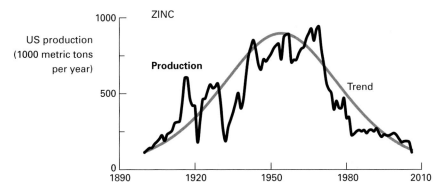

Figure 4.12 US zinc production and fit of a logistic curve to the production data. (Data: US Geological Survey – USGS)

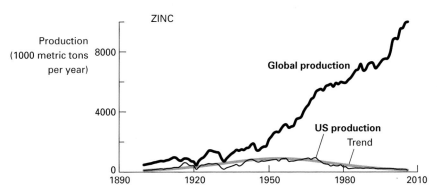

Figure 4.13 Zinc production in the US (lower bell-shaped curve) compared with global zinc production (increasing curve). US production declines as global production increases. (Data: USGS)

The US production of other metals, like iron and lead, has also more or less resembled a bell-shaped curve (Figures 4.14 and 4.15), yet we are not running out of these commodities. In fact, in 2007, US mines produced 52 and 0.12 million metric tons of iron and lead (refined), respectively. The US reserve base of each of these metals is 15 billion and 19 million metric tons, respectively.[29] Total resource estimates (not just reserves) are much higher. For example, the 2007 iron ore resource estimate for the US was 110 billion metric tons and globally it was 800 billion metric tons. Using the highest historical levels of production, the US resources would last over 900 years and the global resources over 400 years. Worldwide there are an estimated 1.5 billion metric tons of lead, or enough to last more than 400 years at the current global production rate. Again, the bell-shaped decline in US production does not suggest depletion of these commodities; rather, it reflects the fact that cheap imports are more economical.

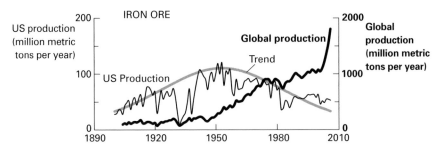

Figure 4.14   US and global production of iron ore. (Data: USGS)

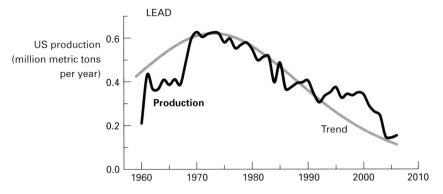

Figure 4.15   US production of lead. (Data: USGS)

The point of the examples discussed above is that the rise, peak, and decline in the production of both renewable and non-renewable resources do not necessarily reflect resource depletion. Rather, production decline can be an indication that more readily available sources exist and price declines have made extraction unprofitable. On a worldwide basis, *none* of the common, globally traded *non-renewable resources* follows the pattern of peak and decline dictated by the Hubbert-type curve of *depletion*. Interestingly, there are plenty of *renewable* biological resources (species) that have peaked and declined due to depletion, even ones that have become extinct. In summary, global data on non-renewable resources do not suggest a pattern of a rise, peak, and decline, and they do not support the notion of global depletion. Of key importance are identifying the factors responsible for production declines and determining whether such declines necessarily reflect exhaustion of the resource.

Generalizing the Debate: Resource Economists versus Neo-Malthusians

The non-renewable resource depletion debate is represented by two camps. On one side is a group of resource economists who maintain that global resource depletion will never occur. Then there are those who are steadfastly convinced that the world is headed for a resource depletion disaster under the pressures of population growth. Although much focus is currently on the future of global oil resources, in 1980, members of the abundance and depletion camps squared off in a debate about the scarcity and destiny of non-renewable mineral commodities. John Tierney of *The New York Times*[30] described a famous bet. On one side was Stanford University professor of biology Paul Ehrlich,[31] and on the other was University of Maryland professor of economics Julian Simon. Ehrlich's 1968 popular book, *Population Bomb*, spoke of an unavoidable, nightmarish future world plagued with catastrophic famine, pollution, and commodity scarcity. In contrast, Simon's 1981 book, *The Ultimate Resource*, optimistically predicted an ever-improving world in terms of human health and longevity, environmental quality, and commodity abundance, all carried forward by technological innovation and efficiency gains.

The two scholars made a bet on the inflation-adjusted market price of metal commodities. Ehrlich's contention was that scarcity would breed price increases, and such increases would provide clear proof that global resource depletion was in progress. Simon maintained that impending resource scarcity was a myth, and the price of any basket of commodities would go down. According to the terms of the bet, Ehrlich was to select a basket of any five metal commodities with a 1980 market price of $200 for each metal (a total of $1,000) that he felt would be more expensive in ten years. Ehrlich would win the bet if the inflation-adjusted price went up, and Simon would win if that price went down. The amount of the wager was the difference between $1,000 and the 1990 price of the basket of metals. It should be noted that, in principle, Simon had uncontrolled risk if the prices went up, but the most that Ehrlich could lose was $1,000 if all of the commodities became worthless by the end of the decade. Ehrlich selected tin, tungsten, chrome, copper, and nickel as the most likely to become scarce.

Simon won. These commodities had become less scarce even though global population increased by 21 percent from 1980 to 1990.[32] The inflation-adjusted average of the metal prices dropped by 57.6 percent during the decade. Ehrlich paid Simon $576.07. Even had there been no adjustment for inflation, Simon would have won the bet, as the average price of the basket dropped by 36 percent. Furthermore, after accounting for inflation, every single metal that Ehrlich picked had declined in price by at least 17 percent (Table 4.1). Although Simon won the bet because the entire basket of metals priced out lower, he actually would have won a series of bets had they been on a metal-by-metal basis.

Ehrlich paid off his debt but remains a firm believer in the proposition that resource scarcity will come soon. Simon offered Ehrlich a rematch, but it seems that one bet was enough for Ehrlich. In 1998, Simon died, but it is probably safe to say that he would recommend taking such a bet again and again. Interestingly, following Simon's lead, *The New York Times* writer John Tierney joined forces with Simon's wife to make a bet on the future price of oil.[33] In 2005, the pair bet a Texas energy industry investment banker, Matthew Simmons, $5,000 that the 2010 annual average oil price, adjusted for inflation, would not rise above $200 per barrel. That price is well above the annual average or even the daily spot price of oil at any time through 2008.

The lesson learned from the results of the Simon–Ehrlich bet is that although it seems intuitive that increasing population pressure must cause prices to permanently increase under conditions of global scarcity, this has not happened for any significant globally traded, non-renewable Earth resource. During the 10 years in which the Simon–Ehrlich wager was in place, technology improved, and some minerals became less desirable as cheaper substitutes emerged. Consequently, prices went down as the trend in demand weakened relative to increasing supply.

Table 4.1 Percent decline in metal prices selected in the Simon–Ehrlich bet

Metal	*1980 to 1990 percent decline in price (adjusted for inflation)*
Tin	−73%
Tungsten	−73%
Copper	−29%
Chrome	−18%
Nickel	−17%

In the long term, the inflation-adjusted price trend of many globally traded, non-renewable commodities has declined even though demand, reflected by increasing consumption, has gone up. However, one of the difficulties with using price as a measure of scarcity is that commodity prices are volatile. The price trend is affected by temporary forces that can hinder extraction and delivery. These are not small forces. They range from worker strikes to governmental policy changes to trade agreements, as well as to political unrest and wars. In addition to supply disruptions, there can be unanticipated spikes in demand that cause prices to increase. From this angle, Simon was somewhat lucky, as there have been decade-long periods over which the inflation-adjusted price of a particular commodity has not declined, even though, over the long haul, the price trend has been down. For example, the inflation-adjusted price of tin in the 1970s averaged 54 percent more than in the 1960s and 79 percent more than in the 1950s. The inflation-adjusted price of chromium was higher in 1975 than during any previous time dating back to at least 1950. These metals were part of the Simon–Ehrlich bet. Because metal commodities are mined and processed, their prices are tied to energy and oil costs. So, oil price volatility can enter into the metal price trend. For these reasons, taking either side of such a bet is risky unless the measure of price increase or decrease reflects the very long-term, inflation-adjusted price trend of the commodities.

Figure 4.16 shows three examples of the price and demand for non-renewable resources. The first is zinc, which was discussed earlier. The price has declined by about half since 1900, while global production has climbed tenfold. The second example is aluminum ore, the price of which has dropped by more than half since 1900, while global production has increased 100-fold. Both of these metals are certainly finite, but there is no evidence of scarcity in the global market. The third commodity shown is crushed stone produced in the US. This is a commodity with such a low value that it is not typically imported or exported, but, nonetheless, its price has dropped by half as demand has soared tenfold in the past 100 years. The point is that increasing demand for a finite resource, as reflected by increasing production, does not necessarily create either economic scarcity or price increases.

Copper, the first metal to come into widespread use on a large scale, is a good example of a commodity whose scarcity has been improperly projected. After the invention of the telephone in 1887, copper became an essential commodity in the industrialized world. Demand grew by almost 6 percent a year through the mid-1900s, reflecting copper's widespread use in construction and industry. In 1950, the US Geological Survey estimated worldwide reserves at 91 million metric tons, an amount that only would have lasted for 38 years at the production rates of the day. Copper prices peaked

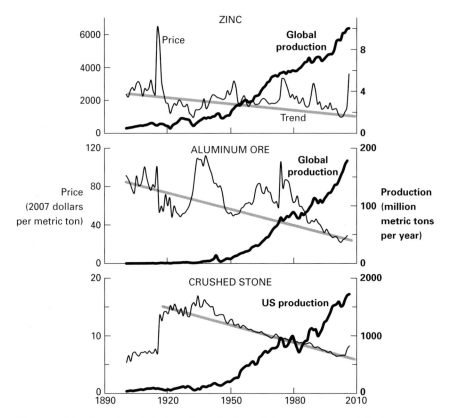

Figure 4.16   Over time, inflation-adjusted prices of the above commodities have declined even though production has greatly increased over time. (Data: USGS)

(temporarily) in the early 1970s as a consequence of demand increases during the Vietnam War. There was concern that the US would run short as the demand for copper continued to increase. Over time, copper wires had been snaked through the US and much of the developed world. But where would the copper come from to meet the demand of developing nations? In 1970, Harrison Brown, who was elected to the US National Academy of Sciences in the same year as Hubbert (1955), wrote a major article in *Scientific American* claiming that even with new discoveries, demand would increase, and the world would deplete its copper reserves by 2001, with lead, zinc, tin, gold, silver and platinum also being depleted during the late 1980s.[34]

Thirty years later, we can see that, clearly, the world did not run out of copper or any of the other metals mentioned in Brown's article. First, there was more copper than estimated. From 1950 to 2006, worldwide cumulative production exceeded 400 million metric tons, almost four-and-a-half times the global reserves estimated by the USGS in 1950.[35] In addition, the USGS in 2008 estimated 550 million metric tons of copper reserves still remained in the world, or six times the 1950 estimate. US copper production has declined by about 30 percent since 1973, but not because there is a lack of resources. There are 3 billion metric tons of copper in land-based deposits worldwide compared to 15.7 million metric tons produced in 2008, about 200 years of availability at the current rate of global production. Second, projected demand for telephone wires did not follow past trends. Fiber-optic technology sprang up and diminished the demand for copper wire. Plastic pipes began to substitute for copper ones. Aluminum substituted for copper in car radiators and power cables.[36]

Since reaching a near all-time high price in 1974, the inflation-adjusted price of copper had dropped steadily by more than half by 2004. However, the price rebounded suddenly during the 2000 to 2010 decade to a near record level. Does this suggest global copper depletion? No. As with most commodities, copper prices are volatile and respond to world events that affect supply and demand. Copper remains an important material and there has continued to be strong demand. In 2007, a sudden 37 percent increase in copper consumption by China coupled with worker strikes in major copper-exporting countries (Chile and Peru) caused prices to rebound. This, like other spikes in demand, might be curtailed by increased supply, new technology, substitution, and decreased demand.[37] In fact, in the early stage of the 2008 global financial crisis, the price of copper on the London Metal Exchange fell from $4.07 per pound in April 2008 to $1.70 per pound in October 2008.

This brings up a key question concerning resource utilization: Do we need any particular commodity? A knee-jerk response is that of course we need metals and fossil fuels. But consider copper. We certainly need the services that copper provides, but do we necessarily need copper? For example, we need to communicate over long distances, and if that need could only be met by telephone lines containing copper wire, then yes, we would have to have copper. However, if our need to communicate can be met by fiber-optic lines or cellular phones, then we don't need copper for that purpose. Our end-use need is communication, not copper. Similar examples can be found for other commodities once thought to be essential.

A few remarks are in order about the nature of production declines over time and substitution and abandonment of resources thought to be essential. Technological changes and environmental concerns can be sudden, and both can result in rapid resource substitution. Production declines can be rapid because substitution of new technology can occur very quickly. There are many examples of common items for which substitutions have been prompted by technological innovations or environmental concerns. At the turn of the twentieth century, the world stopped needing horses for transportation. In just a matter of decades, automobiles became widely available. In addition, the world rapidly switched from fuels like whale oil in the 1850s to kerosene in the 1860s.

An example of substitution for environmental reasons is mercury, an environmental toxin that has made its way into the food chain. Mercury damages the human immune, enzyme, genetic, and nervous systems. Its production in the US plummeted rapidly from 1,060 metric tons (15 percent of global production) in 1980 to zero in 1993, and no production in the US occurred thereafter. Since 1948, the global production of mercury has followed a bell-shaped curve with a peak in 1971, yet there is plenty of mercury left (Figure 4.17). Worldwide, known resources total about 600,000 metric tons, which is about a 60-year supply at the 1971 peak use rate and an over 400-year supply at rates produced since 2000. Why has global mercury use declined so dramatically? Substitution. Mercury-zinc batteries have been replaced by lithium, nickel-cadmium, and zinc-based batteries. Mercury thermometers have been largely replaced by digital thermometers or by thermometers containing an alloy of gallium, indium, and tin. Mercury in fluorescent lights is being replaced by light-emitting diodes (LEDs) that contain indium. Dental fillings that once used mercury as part of gold and silver amalgams are now made of ceramic, and mercury used as a fungicide in latex paint has been replaced by organic compounds.[38] Generally speaking, we really do not need mercury even if we do need the services that it once provided.

Mercury is interesting because it is one of the few globally traded, non-renewable commodities that has gone through a complete, and perhaps permanent, cycle of production rise and decline. Its global production rise, peak, and decline are similar to other regionally produced commodities for which there has been substitution. In the case of mercury, substitution was driven by its toxicity and cost to human health even when in low concentrations. However, for many commodities, substitution has been prompted by high economic cost. Lower prices stimulated higher and higher demand through the early 1960s (Figure 4.17). Peak production of mercury in 1971 was accompanied by higher prices, thereby spawning innovative technology and substitution of other commodities for the many uses once reserved for mercury. This substitution resulted in lower and lower demand, coupled with a collapse in both the price and production.[39]

Importantly, technological advances can create a leapfrog effect, where some commodities become obsolete before they are adopted, and thus some consumers skip their use altogether. For example, cellular phone technology is increasing at such a rate that many parts of the world will likely never see

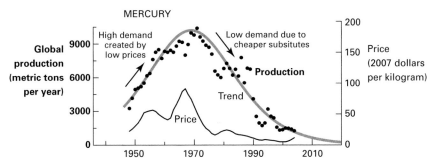

Figure 4.17 Global mercury production and price declined once it was appreciated that it was a toxin, even in low concentrations. In the US, mercury has not been produced since 1993. Price shown is the five-year running average. (Data: USGS)

telephone poles and wires because a technological jump to wireless systems has made wires obsolete. In remote locations in islands of Indonesia, mobile phones are common and landlines were never part of the technological landscape. In many cases, the use of a resource for particular purposes has been rapidly abandoned in favor of superior substitutes. Ironically, such leapfrogging can, at least temporarily, leave developed regions with lower-level technologies than underdeveloped regions, which manage to skip entire generations of outmoded gadgets.

The concept of end-use services goes far beyond the lessons learned about impending copper depletion, a widely held belief in the 1970s. For example, does the world need all of the jet fuel that it consumes? In the context of business, although it is valuable to travel and have face-to-face meetings, our end-use needs are often communication, interaction, and getting things done. Imagine a communications system that produced images and sound quality that were far more realistic than those provided by our common telephones or personal computer video calls and conferences: a three-dimensional, immersive, communication environment that had extraordinary surround-sound. If long-distance interactions were almost as pleasant, real, and productive as face-to-face meetings, then perhaps this new communication system could substitute for at least some share of business travel. The underlying business need in this example is communication, not travel, and certainly not jet fuel. This is not to suggest that travel should cease but, in many circumstances, business travel could be reduced with minimal sacrifice. It is even possible that more frequent, easy, and inexpensive communication might enhance certain interactions and thus reduce the need for travel. At the same time, resource depletion, pollution, and risks could be diminished. Whether or not the ease and frequency of inexpensive remote contact is a desirable substitute for face-to-face meetings is a value judgment, but substitution of one means of communication for another is clearly possible and can result in reduced use of a resource, such as jet-fuel produced from oil, that is thought to be essential.

Back to Oil

So what about oil? Is there evidence of scarcity? Clearly, we cannot equate patterns of consumption of oil with those of renewable resources like tobacco or flaxseed. By definition, unless totally eliminated, renewable resources can be more or less readily replenished. On the other hand, once oil is consumed, it is gone forever: it is indeed a non-renewable resource.

The question before us is whether oil will follow a pattern of production similar to that of other non-renewable commodities, like the metals. What are the factors that have shaped the historical pattern of oil production, both locally and globally? We can begin to answer this question by returning to the data. As pointed out earlier, the global trend of oil production has not followed the Hubbert-type curve (Figure 4.18).

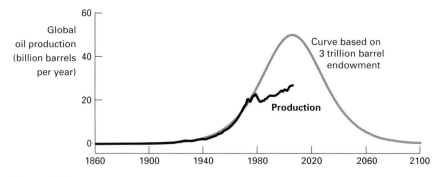

Figure 4.18   Global oil production data fit of a logistic (Hubbert-type) curve to production data through 1980 using the USGS estimate of 3 trillion barrels for the global oil endowment. (Data: EIA)

Using the 3 trillion barrel value for the global oil endowment from the USGS Assessment, Figure 4.18 shows the good fit of the Hubbert curve to the production data through 1980. After that, there was a brief marked decline in production followed by a straight-line increase. The Hubbert curve indicates that peak global production should be about 50 billion barrels per year. Data show that fewer than 30 billion barrels per year are currently produced, a value far beneath the peak of this Hubbert curve. The exponential rise that occurred for the 120 years after the first oil wells were drilled in the 1860s has not continued.

Departing from the exponential trend of the Hubbert curve through 1980, the actual trend of oil production has since been replaced by a linear one that has been fairly consistent over that last 26 years. It turns out that oil production from 1983 through 2008 has grown simply in proportion to the global population increase.

As shown in Figure 4.19, for over two decades, there have been about 4.1 barrels of oil produced each year for each person on the planet, or, alternatively, 4.6 barrels consumed per person per year. Those annual per-capita figures have remained essentially constant. Had you wished to estimate annual global oil production from 1983 to 2009, you would have been wrong using Hubbert's curve. You would have been right if you had simply taken the projected population and multiplied it by 4.1; the average absolute error in your prediction would have been less than 2 percent. In the past 26 years, both world population and global oil production have increased by about 40 percent. It only takes simple math and the right estimate of population growth. Projecting to the year 2075, when the world population is expected to be at its maximum, and using a United Nations estimate of the world population maximum of 9.2 billion people, one can estimate (multiplying 9.2 billion by 4.1) a future production value of just under 38 billion barrels per year.[40] That is, 38 billion barrels per year would be the peak in production in about 65 years (i.e.,

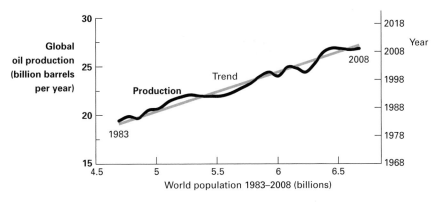

Figure 4.19 Global oil production vs. population showing that for the past 26 years, approximately 4.1 barrels per person per year have been produced. Years shown on the right axis apply approximately to the population trend shown. Note that oil production values do not include processing gains and are not the same as consumption values, which are higher. (Data: oil production, EIA; population, Economic Research Service, USDA)

2075) when the world has over one-third more people. This is beyond the many estimated dates of peak oil. This estimate of oil production would still fall short of the anticipated peak-oil value of 50 billion barrels per year given by the Hubbert curve fit to data through 1980 (Figure 4.18). For reasons dis-cussed later, the global annual production rate may be substantially less than 38 billion barrels, but not necessarily because oil reserves will be depleted.

Returning to the US oil production data, there is an explanation for the timing and immediate causes of US peak oil that is not related to production and consumption. Although the US is clearly a mature oil-producing region, and most readily accessible oil has been discovered, the US has twice as much technically recoverable oil than suggested by historical estimates of the oil endowment. Production today, beyond the peak, has not fallen off as rapidly as predicted by Hubbert and his disciples, and the prospects are promising for the US to continue to be a significant oil producer (currently third in the world). Of course, US oil has become more challenging to produce as the easily discovered and extracted oil has been depleted. However, oil produc-tion remains profitable in the US. Interestingly, followers of Hubbert pre-dicted in 1981 that energy costs of drilling and extraction would grow so high sometime between 2000 and 2004 that the effort to exploit domestic oil resources would not be worth the return on investment.[41] That situation did not occur.

One reason contributing to the decline in US oil production in the 1970s was cheap imports. From 1955 to 1970, imports accounted for about 10 percent of US oil consumption, but this value had doubled by 1973 and tripled during the following five years (Figure 4.20).[42] What is the relationship between inexpensive imports and the decline in US domestic production? Can it be assumed that the magnitude and timing of imports occurred because the US had largely drained its domestic supplies as Hubbert suggested, or did the availability of a cheaper alternative play a role?

In his initial work in the 1950s, Hubbert predicted that peak oil would occur in 1965. At that time, the Middle East was known to have abundant, easily extractable, and inexpensive oil. However, it was problematic at that time to transport significant quantities of Persian Gulf oil to the US market. Until 1965, the capacity of the vast majority (92 percent) of oil tankers was 60,000 tons or less (about one-third to one-fifth the size of the 200,000–300,000-ton supertankers to come). Oil had to be shipped in small tankers through the narrow Suez Canal to the Mediterranean Sea and out across the Atlantic Ocean to the US. For companies hoping to import oil to the US, this was a major obstacle.

World events compounded the logistical problems of importing oil from the Middle East. In 1967, Egypt closed the Suez Canal after its six-day war with Israel, and the canal remained closed until 1975 – years after the 1970 peak in US oil production. Given that situation, shipping large quantities of oil to the US was both infeasible and uneconomical. The technological development of supertankers, or Very Large Crude Carriers (VLCCs), solved the problem (Figure 4.20). In 1967, only three such commercial tankers existed in the entire world. By 1970, there were 133 supertankers, including six over 300,000 tons.[43] None of these supertankers could fit through the undersized Suez Canal. However, given their enormous payload, it was cheaper to ship Persian Gulf oil in supertankers around the tip of Africa to Western Europe and North America than to have many smaller tankers traveling the much shorter route from the Red Sea through the Suez Canal, which then was closed anyway, and finally through the Mediterranean and across the Atlantic.[44]

It was at least in part the technological innovation of the supertanker that spawned an increase in imports of inexpensive Persian Gulf oil to the US after 1971. From 1965 to 1970, worldwide tanker capacity grew by 76 percent, and the total tonnage of tankers "on order" increased by 118 percent (432 supertankers on order) in 1970. Perhaps not coincidentally, this is when peak oil occurred and US production declined. Hubbert considered oil depletion as the sole driving force in his model of peak oil. There was no room

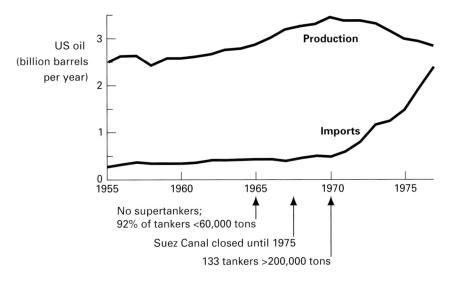

Figure 4.20 Decline in US production and increase in imports corresponded to the development of the supertanker, which made Persian Gulf imports feasible by 1970. (Data: oil production and imports, EIA; tanker values from Rifai (1974)[45])

for other mechanisms responsible for the timing of peak oil, such as the relationship between expensive US domestic production and inexpensive Persian Gulf imports, undersized cargo lanes, shipping technology of the day, and conflicts in the Middle East. The infeasibility in 1965 of importing oil to the US meant that Hubbert's projected decline in US production beginning in 1965 would have left the US without sufficient oil. The 1965 peak that Hubbert predicted, followed by insufficient supplies, never happened.

As Edward Porter, Research Manager at the American Petroleum Institute in 1995, wrote in his marvelous analysis of US and global oil depletion:

> The decline in U.S. supply after 1970 did not indicate that the U.S. was "running out" of oil, but rather that the costs associated with much of remaining Lower 48 resources was no longer competitive with imports from lower cost sources worldwide. Consequently, the decline in U.S. supply after 1970 represented not a signal of growing global resource scarcity, but rather a signal of growing global resource abundance.[46]

Porter shows that the incremental cost of oil production in 1994 was 25 to 30 times higher in the US than in Saudi Arabia, and presumably an analogous differential existed in 1970.[47]

In the coterminous US, the easily accessed, conventional oil resources are being played out, but the declining trajectory of oil production is not fixed. The decline in oil production from the lower 48 states is likely to continue, although there is nothing requiring the decline to mirror the increase in production. During the period of decline since 1970, additional discoveries in Alaska and the Gulf of Mexico, as well as reserve growth, have helped maintain much of the US domestic supply. The addition of unconventional oil resources, as discussed later, may further stem the decline in the US.

In the case of non-renewable resources, the inflation-adjusted price of an abundant commodity would be expected to decline over time. On the flip side of this relationship, the inflation-adjusted price trend should increase for a scarce commodity – one mined to near exhaustion. Long-term oil prices have varied tremendously over the past 150 years or so. But if prices are adjusted for inflation, there has been no long-term upward price trend as one would expect if indeed oil were becoming scarcer (Figure 4.21). If the price of oil is considered relative to worker wages,[48] rather than adjusted with respect to the consumer price index, the long-term price has declined over the past 150 years or so (Figure 4.22). The price of oil always has been volatile. The price has dropped dramatically after each price spike, which one would not expect if the commodity were growing scarce.

Even with the caveat about the lack of a free and undisrupted oil market, the inflation-adjusted price of oil has followed a flattening trend over the past

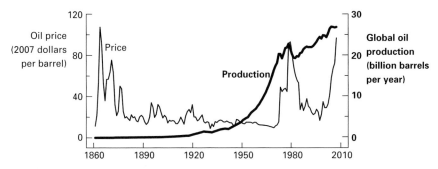

Figure 4.21 Inflation-adjusted average annual oil price and global oil production over time.[49] (Data: oil production, EIA; oil price, *BP Statistical Review*)

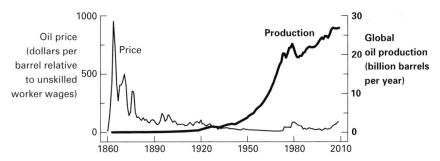

Figure 4.22 Average annual oil price relative to unskilled worker wages and global oil production over time. (Data: oil production, EIA; oil price, *BP Statistical Review*; wage deflator, Williamson (2008)[50])

100 years with the exception of price spikes (and consequent panics) that have nothing to do with actual oil depletion. There have been temporary price upswings associated with world events and OPEC's control on production. For example, the price of oil jumped in 1973 due to the OPEC oil embargo following the Yom Kippur (Arab–Israeli) war. The price climbed again in 1978–80 due to the Iranian revolution and the Iran–Iraq war. In 2003, it again rose due to an oil-worker strike in Venezuela, then the world's fifth-largest oil exporter and the source of 12 percent of US imports. Adjusted for inflation, the 2008 average annual price of oil at $94 rose above $93 per barrel to exceed its average price in 1980 (see Table 4.2). If oil prices climb in the years to come, demand for oil will decrease as alternatives emerge.

The price of oil is far more abstract to most of us than the price of gasoline. The price of gasoline has also followed a declining trend over the past century. Mirroring the inflation-adjusted price of oil, the dips in the

Table 4.2   Nominal and inflation-adjusted annual average oil price per barrel since 1864

Year	Nominal $/barrel	2007$/barrel
1864	8.06	106.05
1870	3.86	61.27
1880	0.95	19.39
1890	0.87	19.77
1900	1.19	29.02
1910	0.61	13.26
1920	3.07	31.98
1930	1.19	14.69
1940	1.02	15.00
1950	1.71	14.74
1960	1.90	13.29
1970	1.80	9.63
1980	36.83	92.77
1990	23.73	37.67
2000	28.50	34.30
2005	54.52	57.88
2006	65.14	67.02
2007	72.39	72.39
2008	97.26	93.70

Data: Nominal prices, 1861–1944 US average, 1945–83 Arabian Light, 1984–2008 Brent, BP Statistical Review; Inflation-adjustment (Sahr, 2009)

inflation-adjusted prices of gasoline have reached new lows, yet the low prices have been interrupted by price spikes (Figure 4.23). The price spike in 2008 drove prices above historical levels. Given abundant known global oil reserves and the fact that, at most, one-third of the global oil endowment has been exhausted, it is difficult to argue that elevated prices during 2005 to 2008 indicate global oil depletion. For those who believe that abundant oil resources remain, if history is a guide, the long-term (over decades), inflation-adjusted trend should favor lower oil prices than the elevated values that occurred in 2008. With the decrease in global demand of late 2008 and 2009, indeed the prices of oil and gasoline fell to levels consistent with their historical trends.

Scarcity Rent

For manufactured commodities, prices reflect the cost of production. However, in the case of non-renewable resources, prices also reflect the scarcity of the

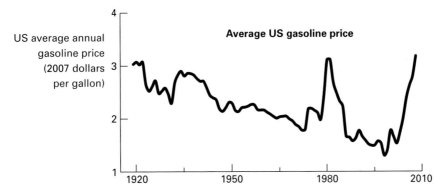

Figure 4.23 Inflation-adjusted annual average US gasoline prices over time. Note that the gasoline prices shown are annual averages and do not indicate variations that have occurred during any particular year. The price decline that began in mid-2008 continued into early 2009. (Data: gasoline price, EIA; inflation-adjustment, Sahr (2009)[51])

commodity itself. If a resource is being mined toward exhaustion, its price in excess of production cost will increase with time as it becomes scarcer. So the owner of the non-renewable resource will show no preference between mining the reserve and leaving it as it was in the ground where its value will grow. The unit price of the remaining reserves should increase at the same rate that the owner would otherwise earn by selling the mined commodity and investing the proceeds. The classic "Hotelling" economic theory predicts that market forces will place a "scarcity rent" on the price of an exhaustible resource. **Scarcity rent**, the difference between the price and the marginal (highest) cost of production, is a measure of this intrinsic economic property of non-renewable resource exhaustion. According to many economists, for a commodity that is being "mined out," or exhausted, an increase in price relative to marginal extraction cost will grow with time. High scarcity rent suggests exhaustion of reserves,[52] and low scarcity rent indicates that depletion of a non-renewable resource is not on the horizon.

Evaluation of the scarcity rent of oil does not hint that oil is scarce in an economic sense. The downward long-term price trend of oil is not consistent with an increase over time in scarcity rent. The price of oil, as echoed in the inflation-adjusted price of gasoline (Figure 4.23), generally decreased between 1920 and 1975, increased briefly between 1975 and 1980 (due to OPEC), and declined again between 1980 and 2002. As noted about oil by Professor James Hamilton of the University of California, San Diego, "As a result, economists often think of oil prices as historically having been influenced little or none at all by the issue of exhaustibility." Prices did increase from 2002 through mid-2008, but Hamilton also noted that the declining price of crude oil futures contracts sold in October 2007 with maturity dates to 2015, which is inconsistent with increasing scarcity.[53]

From 2004 to 2007, the cost of new oil production in the offshore US climbed to about $60 per barrel. This might be interpreted as a sign that the world is straining to obtain every bit of oil. The fact is that very little oil is produced at the highest cost. In 2008, 44 percent of the global supply was provided by OPEC, and very low-cost OPEC producers in the Middle East have an oil extraction cost of about $2 per barrel. Their main producer, Saudi Arabia, has maintained spare production capacity (at a profit) when oil demand has weakened and oil prices have fallen. It is the long-term plans and production strategy of oil-rich countries in the Middle East, with their enormous reserves, that largely determine the cost at which supply meets demand during times of global economic growth. Furthermore, with regard to higher-cost producers, over time technological advances have lowered the unit cost of production of many non-renewable resources (e.g., oil), even though the "quality" of the remaining reserves has diminished or reserves in more challenging environments have been targeted for exploitation. The important point is that the downward price trend, even in the face of cartel oil-production limits and new high-cost finding and production technology, is among the best indications that oil resources are not being exhausted.[54,55,56]

Myth III: Resource Assessments Provide Useful Endowment Estimates

Peter McCabe, formerly a scientist with the US Geological Survey in Denver, Colorado, for 20 years, is an expert in the analysis of global oil resources. Perhaps more than any other analysis of oil resources, his stands out as remarkably instructive. McCabe analyzed historical estimates of the oil endowment, made from 1949 to 1996 by the USGS and other reliable sources. The estimates showed a distinct trend.[57] McCabe's results for the US are displayed in Figure 4.24, with the addition of post-1996 data that have become available since his work.[58] The sequential assessments conducted over the past 50 years provide various estimates of the endowment of US crude oil. Not only do the sequential historical estimates differ, but they show a systematic increase from the first assessment, made in 1948, of 110 billion barrels through the 1996 USGS estimate of over 330 billion barrels, and to the 2006 DOE estimate of 398 billion barrels of conventional US oil. Over time, assessments have added about 5 billion barrels per year to the estimated US oil endowment. In its day, each assessment was believed to be more accurate and more credible than the last. In reality, there was no more oil during one period than another, but sequential assessment values consistently increased.

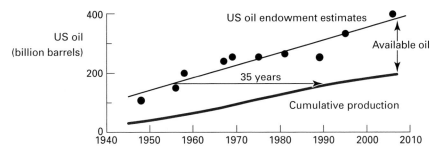

Figure 4.24   Cumulative oil production in the US and sequential assessments show a systematic increase in the US oil endowment. Oil identified in any assessment is produced over about the following 35 years (after McCabe (1998)[59]). (Data: oil production, EIA; endowment value since 1996 added based on DOE (2006)[60])

Also shown in Figure 4.24 is cumulative production over time, which roughly parallels the increase in the estimated US oil endowment. The difference between the endowment (presumably, what we have) and cumulative production (what we have used) is the **available oil**. Comparing the historical cumulative production to the endowment, the interesting thing to note is that the known volume of available oil has consistently been projected to last about 35 years into the future. For example, the US oil endowment estimate in 1955 was about 150 billion barrels, and yet it was not until 1990 that this much oil had been produced. There were no signs of complete depletion in 1990, even though the 1955 endowment was exhausted. In a similar analysis, McCabe showed that US reserve estimates have also increased with sequential assessments. At any time over the past 80 years, the estimated oil reserves were projected to be sufficient to meet US demand for 10 to 15 years. With each successive assessment, the projection of reserve depletion has been pushed out another 10 to 15 years.

McCabe also inspected historical assessments of global oil resources. Figure 4.25 shows that assessments of the worldwide oil endowment climbed from 0.6 trillion barrels in 1948 to over 3 trillion in the last USGS Assessment. Assessments have increased the endowment value by about 35 billion barrels per year, which is greater than the highest annual global oil production rate. The available oil has remained steady and perhaps even increased slightly since data were first reported. Consistent with McCabe's evaluation of oil endowment data for the US, the available oil for the entire world is sufficient to accommodate production for 40 to 50 years. If anything, the latest USGS Assessment suggests that the available oil will last even more than 40 to 50 years.

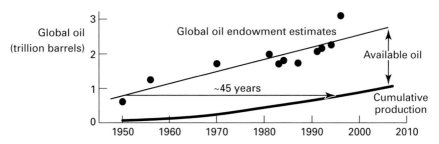

Figure 4.25   Assessed global oil endowment values have increased with time, providing for 40 to 50 years of oil production. All estimates were made by the USGS (see McCabe (1998)[61]) with the exception of early estimates by Weeks (1948)[62] and Hubbert (1956)[63]. The value in 1969 is the average of two values made by Hubbert when working for the USGS. Cumulative production from Salvador (2005)[64] through 1989 and EIA from 1990 on. (Figure after McCabe (1998))

There are two important implications of McCabe's global oil results. First, it is remarkable that sequential assessments seem to consistently find enough oil to last the world another 40 to 50 years. It is as if our assessment window into the future only enables us to see that far, whether we evaluate the state of oil resources in 1950, 1970, or today. Second, the fact that the endowment value is not a constant contradicts the application of Hubbert's logistic curve approach. Recall that the area under the Hubbert curve corresponds to the oil endowment. Hubbert's approach is aimed at determining how long it will take to deplete that fixed quantity of oil. If sequential assessments find more and more oil, the oil endowment is not a known, fixed quantity, and Hubbert's conceptual model of depletion was wrong. Hubbert neglected the fact that sequential assessments systematically report greater available oil resources. Over time, Hubbert struggled with this fact. He revised his estimates of oil depletion, generally relying on higher and higher endowment values, each of which was (and is) the toughest value to acquire.

As unintuitive and annoying as it might be, the value of the supposedly fixed endowment actually follows an increasing trend with successive assessments. The consistent increases in the assessed oil endowment mean that we never know how much oil actually is available. This is the case even though we use more and more sophisticated methods in sequential assessments. Our confidence is high that "this time we got it all," but each new and improved value is ephemeral. Based on inspection of the historical record, any given assessment underestimates the global oil endowment. A snapshot endowment value is sufficient only to compute a lower bound on the time-frame to exhaust the presumed fixed global endowment, because updates suggest that there is always more oil than we thought. The fundamentally incorrect assumption of the Hubbert-curve approach applied to the entire world is that a known mass balance exists and the area under the curve is a quantity that does not change over time.

How can McCabe's 40- to 50-year consumption phenomenon be explained? There are two parts to the explanation. First, as the technology to recover oil improves, so does our ability to estimate its presence. New technologies for both discovery and recovery consistently add to the amount of available oil, so the estimated oil endowment increases with time. Some breakthroughs have included the use of 2D, 3D, and 4D (over space and time) seismic data, new computer simulation models of oil reservoir behavior, and advances in drilling, such as horizontal drilling. Oil that used to remain trapped in the ground can be flushed out with enhanced recovery techniques, so this oil is added to the assessed quantity. In addition, unconventional oil, such as oil sands and heavy oil, which may be completely immobile and requires extra effort to extract, is added to the amount that can be recovered. Economic incentives drive the development of more sophisticated technology to find new accumulations or recover more oil that was thought to be unrecoverable. As price increases, even though that increase might be temporary, new technology comes on line, and more oil is added to the stock. Discovery technologies that prove to be efficient typically are not abandoned if the price of oil goes down.

The second factor that explains McCabe's 40- to 50-year oil consumption time-frame is more subtle but could probably be called the "we have enough not to worry" phenomenon. Assessed endowment values represent a volume that is adequate to meet demand for a specified period of time. It doesn't pay to search for and identify new sources when current availability is sufficient for the next half century. By analogy, when you buy 10 cans of soup and use them up in one month, this does not mean that your consumption of 10 cans per month will lead to a disastrous depletion of soup in your household.[65] You can and will seek more as the amount stored in your pantry diminishes. It is not likely that you would stock up with years of soup even if the shelf life of the cans allowed it. The amount of unopened soup cans that you keep in stock reflects the reserves you wish to maintain to meet your anticipated demand over a reasonable period. It is not cost-effective in terms of storage space or household cash flow for you to maintain too much soup, especially since you are confident that you can readily obtain more. Oil assessments work the same way. Anticipated economic conditions (like household cash flow) and technological capability (like shelf storage space) determine the quantity of a commodity that one believes is available to restock for the future.

Part of the difficulty with oil assessments is that the terms "resource" and "reserve" are not clearly defined, leading to subjective estimates. For both the mined mineral deposits and extracted oil and energy commodities, a reserve is the known material that is technically and profitably recoverable at the time of determination. A resource includes yet-to-be-discovered material as well as known accumulations that currently cannot be recovered. It is difficult to imagine two less useful terms. To quantify a reserve, one must not only locate the oil but also define and freeze both current economic conditions and the state of technology for extraction. Any snapshot of oil recovery price and technology is tremendously variable and uncertain over time. Prices are subject to market forces and world events. As has occurred during the past several years, over short periods, oil prices can swing up 300 percent or drop by 75 percent. On economic grounds alone, oil deep beneath the sea floor or from an expensive, enhanced-recovery operation could be profitable one year and unprofitable the next. When prices fall, the inventory of oil is not lost, but it can become uneconomic to produce. In principle, a series of instantaneous assessments made under conditions of fluctuating prices might categorize oil as a resource one year, a reserve a year later, and then a mere resource again the next year. In addition, technology jumps can be developed and deployed rather rapidly, which can change a resource to a reserve. Consider Canada: reserve estimates made in 2002 were increased in 2005 by a factor of 35 because of new oil-sand recovery technology that had become available. With the new technology, Canadian oil sands can be mined as long as the price of oil remains high or if the recovery technology becomes cheaper. Given today's technology, if the price of oil collapsed, it is imaginable that some Canadian oil sands could not be produced profitably and would no longer be categorized as a reserve (oil that is profitable to extract).

As mentioned above, the definition of resource includes oil that is undiscovered or unprofitable to recover. It is too much to ask of geologists and engineers making resource assessments to speculate accurately about potential locations and amounts of oil that there might be under unknown and unforeseeable economic and technological conditions. All that can be done is to make some assumptions about the nature of technology and price that might exist within some reasonable time-frame and then to estimate the locations and quantities of oil accumulations. This is what was done by the USGS in its most recent Assessment (2000). The researchers picked a time-frame of 30 years, starting in 1996, as the most distant time that they could foresee the state of recovery technology and based their estimate of the endowment on that anticipated state. It is no surprise that in the history of oil resource assessments, we only see ahead 40 to 50 years of likely consumption. Beyond that time-frame, demand, economic conditions, and technology cannot be forecast with any reasonable certainty.

Much of the above analysis of the "time to deplete the available oil" considered the oil endowment, that is, all available oil. A similar picture emerges if one inspects oil reserves (oil that is technically and profitably recoverable at the time of determination). Annual estimates of global oil reserves, reflecting a much smaller volume than the total available oil resource, have consistently increased with time, year after year, as shown in Figure 4.26. In fact, global reserve estimates have doubled over the past 25 years. Since 1980, the ratio of each year's estimated oil reserves to the consumption rate in the year of each assessment has increased from 28 to 43. Yet there is seemingly more concern now than ever before that we are running out of oil. A snapshot assessment in 1980 would have suggested that oil reserves would have been gone by 2007, just 28 years later. But in 2008, there remained 43 years of reserves, even though demand had increased. Over the past 25 years, it appears that the world has found it sufficient to maintain a secure stock of oil reserves capable of meeting global needs for, on average, about 35 years. That appears to be enough soup on the pantry shelf.

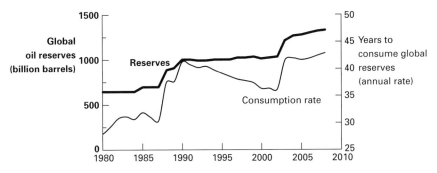

Figure 4.26 Global oil reserves and the number of years to consume those reserves at the rate of consumption in each year have both increased over the past 25 years.[66] (Data: EIA)

The Missing Mass Balance

The issue of resource assessments as a means of quantifying the oil endowment in perpetuity brings us to a fundamental problem with the approach for prediction of oil production used by Hubbert and others. As noted earlier, the most appealing scientific attribute of the Hubbert-curve approach to predicting oil production is that it is based on a mass balance. All oil resources are quantified, and then the approach determines the rate at which they are produced and the time-frame over which they will last. The idea is plain and simple. Unfortunately, the application of a mass balance is the weakest aspect of the approach in that the value of the oil endowment has been nothing but inconsistent. Hubbert himself repeatedly revised his estimates of the oil endowment. For example, in his first forecast, made in 1956, he adopted 1.25 trillion barrels for the global oil endowment, while his last forecast used a range from 1.6 to 2.1 trillion barrels.[67] This is still lower than the USGS 2000 Assessment value of over 3 trillion barrels.

The greatest modern proponent of the concept of peak oil has been Colin Campbell, who founded the Association for the Study of Peak Oil. Campbell has predicted the peak and downturn in global oil production time after time. He first claimed global peak oil production would occur in 1989, based on an oil endowment figure of 1.58 trillion barrels. Campbell produced various increasing estimates of the oil endowment figure in 1990, 1995, and 1996.[68] In a 1998 article in *Scientific American*, Campbell and Laherrère estimated the oil endowment at 1.8 trillion barrels and showed world oil production peaking by 2003 at 26 billion barrels per year.[69] As the peak failed to materialize, Campbell refigured his endowment value at 1.925 trillion barrels in 2002, stating that only 150 billion barrels of oil remained to be discovered, and set the peak oil production date at 2010.[70] Given that over 1 trillion barrels of oil have already been produced and 1.34 trillion barrels were in reserve in 2009 (*Oil and Gas Journal*, 2009), there is no room for new discoveries or reserves in Campbell's latest oil-endowment estimate.[71] As of 2009, there was no peak in sight due to global oil depletion, and production reached 27 billion barrels per year. The point here is not that Campbell's particular predictions of global oil production have been wrong. However, the method of projecting the peak in global oil production is intrinsically flawed because it is based on an assumed mass that has been elusive and systematically underestimated. It is the process of producing a fixed, low estimate of the global oil endowment, coupled with a projection based on a historical curve fit, that is fundamentally unsound.

Counter-Argument to OPEC and Industry Exaggeration of Reserves

In the late 1980s, there was significant concern that OPEC had exaggerated its estimated reserves, given sudden increases in the values reported by member nations. However, when the USGS compared the reserve growth of giant fields in OPEC versus non-OPEC countries, non-OPEC reserve growth, 63 percent, was nearly three times the reported OPEC reserve growth, 22 percent, from 1981 to 1996.[72] Even if OPEC had inflated its reserve value, their exaggeration paled in comparison to the "unchallenged" estimated rate of increase in reserves of giant fields of its non-OPEC rivals. Although OPEC operates behind a veil of secrecy, its reporting of a jump in reserves is not in itself cause for alarm when considered in the context of even greater non-OPEC reserve growth.

The most notable example of oil-reserve exaggeration in the private sector was that of Royal Dutch Shell in 2004, when it announced a succession of downward revisions (five in all) of the oil reserves it had booked in 2002, totaling a 20 percent reduction in reserves.[73–77] The motivation for exaggerating claims of reserves might have been to bolster executive compensation, or perhaps it had fears of losing bonus payments from Nigeria for new reserves.

There is no evidence that the calculated volume of Shell's oil reserves has changed, and the expectation is that the oil ultimately will be recovered even if there is an admitted delay in production. Shell blamed the revision on lack of "project maturity," which too is undefined but likely has to do with host-country stability and support, business priorities, and lack of 3D information from wells about the extent of an oil field. There is no question that Shell's revisions uncovered a serious problem of accounting and reporting, and corporate misdoing is implied by their settling with the SEC. However, it is not to be inferred from the Shell experience that the world's reserves are in decline and there is less oil now than in 2004. Judging by global estimates, evidence is to the contrary, as global oil reserves increased 32 percent from 2000 to 2009, even with Shell's decreased reserve estimates.[78]

The Shell case raises a larger issue of corporate reporting of "booked reserves" that persists but has little to do with known quantities of exploitable oil. The booking of reserves involves two very different tasks: estimating how much oil exists in a particular field under consideration and sub-classification of those reserves as commercially viable or not. The SEC (Rule 4-10) requires that booked or "proved reserves" be "the estimated quantities of crude oil, natural gas, and natural gas liquids which geological and engineering data demonstrate with **reasonable certainty** to be recoverable in future years from known reservoirs under existing economic and operating conditions, i.e., prices and costs as of the date the estimate is made." By adopting the nebulous definition of geologists, the SEC requires that to be a "booked reserve" there be "reasonable certainty" that oil can be recovered profitably under current conditions, "reasonable certainty" that an economic method exists to bring the oil to market, and "reasonable certainty" that there is a willing and able buyer. This definition suffers from the long-standing problem of subjectively distinguishing a reserve (profitable with current technology) from a resource (existing). The ambiguous distinction is particularly problematic when it comes to corporate reporting. What a particular company defines as a booked reserve depends on its assessment of the certainty of oil occurrence and on the technology and economics of recovery and delivery, all of which are subject to change. In fact, Shell debooked reserves in Kazakhstan while its French partner Total did not. Similarly, Shell debooked Norwegian North Sea reserves, but its partner BP did not follow suit.

The SEC revised its petroleum reserve reporting rules effective in 2010 to reduce uncertainty in disclosures. The changes include: (1) classifying proved reserves based on a 12-month average price rather than a one-day, end-of-year price, (2) authorizing the use of new technologies to define the volume of proven reserves, (3) categorizing coal that would be converted into oil and natural gas as petroleum reserves, (4) defining reasonable uncertainty based on more quantitative methods, and (5) not requiring the disclosure of probable and possible reserves.[79] These revisions should add greater transparency to the reserve-booking process.

Myth IV: After So Much Exploration, There Is Little Oil Left To Be Found

Underlying the oil-depletion debate is the notion that from now on, oil discoveries will be few and far between. The concern about depletion continues even though reserves continue to make new all-time highs – 1.34 trillion barrels (2009, *Oil and Gas Journal*), an increase of 10 billion barrels over 2008 and 25 billion barrels over 2007.[80] Assuming that the global oil endowment figure is known and fixed at about 3 trillion barrels (using the USGS Assessment figure), about one-third of the endowment has been exhausted. Is this cause for concern, given the consistent increases over time in estimated reserves and oil endowment figures? Those who maintain that oil is abundant point to two sources of oil: reserve growth and new discoveries. First we will discuss prospects in the US and then those worldwide.

US Oil: Reserves

Proven oil deposits are estimates and are subject to revision. In the US, reserve values in previously discovered oil fields have increased rather consistently. This increase with time in estimated reserves from previously discovered fields is termed "**reserve growth**" or "field growth." Such growth occurs because initial estimates are conservative; exploration, drilling, and production technology improves over time; and economic factors, such as economies of scale, come into play. For the past decade, 75 percent of new reserves in the US were "discovered" within or in the neighborhood of known fields.[81]

In Figure 4.27, US reserve values over time are subdivided into revisions of reserve estimates, extensions to existing fields, new finds in existing fields, and new field discoveries. Two observations fall out. First, during the past 60 years, only 15 percent of US-produced oil has come from discoveries in areas that were not already identified and under production. Second, additions to reserves through revisions of reserve estimates, field extensions, and new finds in existing areas have produced the equivalent oil volume of a Prudhoe Bay (Alaska) every five to six years over the past 60 years![82] When the field-growth components of US reserves are inspected, the claim that "there is no new oil to be found equal to the volume in Prudhoe Bay" does not mean that similar quantities of oil are not being found regularly, even without massive discoveries in areas such the Gulf of Mexico. Even ignoring more discoveries, there have been consistent "discoveries" of a lot of "old oil."

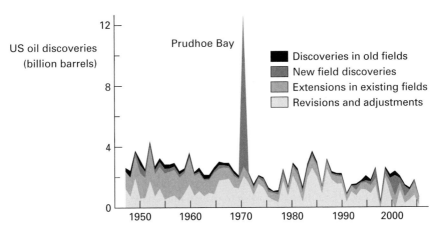

Figure 4.27   The sources of US reserves. Most oil reserves come from reserve growth, with only 15 percent from discoveries in new fields. (Data: from DeGolyer MacNaughton, *20th Century Petroleum Statistics, 2007*; based on API and EIA estimates. Most recent data from EIA. A similar figure with data through 1996 appears in McCabe (1998).[83])

US Oil: Discoveries

Consider the US, which is in many ways a worst-case example for oil prospects, since oil discovery crested in 1935, peak oil production occurred in
1970, and the declining trend in production has lasted over 35 years. So what
is there left to discuss? In the US, discoveries in excess of expectations have
been made since shortly after Hubbert's predictions of oil depletion. The
Prudhoe Bay field, with an estimated 13 billion barrels, was discovered in
1967 and began to produce in 1977. As shown in Figure 4.28, within eight
years, US total oil production rose to within 5 percent of its 1970 peak, at
that time 14 years in the past. Although Hubbert's prediction was based on
the coterminous US, massive oil finds like Prudhoe Bay are not necessarily
anticipated by the Hubbert-curve approach. There have been 14 giant
fields with over 0.5 billion barrels of oil discovered in the US since the
Alaska Prudhoe Bay find (1967), and they sum to over 10 billion barrels.
Furthermore, oil production in the coterminous US has been maintained at
about half of what was produced in the peak production years of 1969 to 1971
– far more than anticipated.

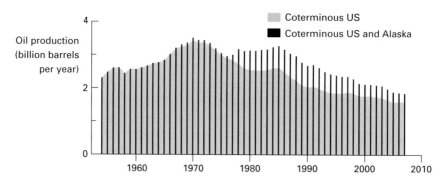

Figure 4.28   US oil production with and without Alaska, showing the rejuvenation in
production following the unanticipated discovery of the Prudhoe Bay oil field. (Data:
EIA)

    The US has other enormous sources of oil that are recoverable with today's
technology, according to estimates of the US Minerals Management Service
for the outer continental shelf (Figures 4.29 and 4.30).[84] In total, this region
is estimated to contain 86 billion barrels of oil and 75 billion barrels of oil
equivalent in natural gas.[85] Over half of these petroleum resources lie beneath
the Gulf of Mexico. Cornell University professor of geology Larry Cathles
estimates that there might be much more oil and gas than that: as much as a

trillion barrels of oil and gas in just a portion of the Gulf sediments, although unconventional recovery methods would be required to produce them.[86]

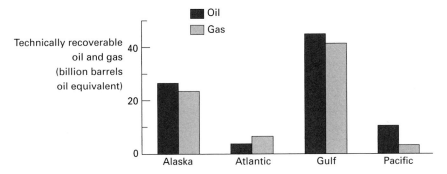

Figure 4.29   Technically recoverable continental shelf US oil and natural gas. (Data: US Minerals Management Service 2006 estimates)

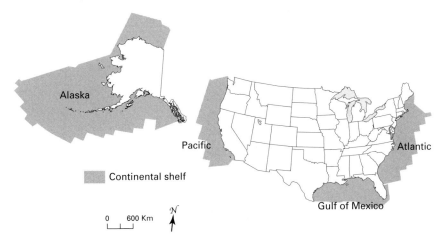

Figure 4.30   Location of continental shelf regions of the US believed to contain rich oil and gas resources. (Based on US Minerals Management Service map)

In 2006, the Department of Energy maintained that there was far more technically recoverable oil in the US than previously estimated. The DOE estimate stands at over 1 trillion barrels of US oil in place with 430 billion barrels of that total being conventional liquid oil or recoverable using enhanced methods. Conventional oil resources in the form of undiscovered oil and reserve growth account for 190 of the 430 billion barrels. Enhanced oil recovery methods are projected to produce another 240 billion barrels.

If we consider the 197 billion barrels produced through 2008 (in the coterminous US and Alaska), the total exceeds 625 billion barrels. To put 430 billion barrels of remaining potential US oil resources in perspective, at the 2008 US rate of production, depletion would not occur for well over 200 years. The implications of the additional identified US resources are twofold. First, peak oil occurred long before half of the US oil endowment had been consumed. In general, even if peak production occurs, there is little reason why the peak for a large region consisting of many oil fields should occur at the halfway point. Second, the various US oil endowment values used by Hubbert of 150, 170, and 200 billion barrels are each a small fraction of the current estimate of the ultimate US oil endowment based on the DOE's estimate of in-place oil (1.25 trillion barrels).[87] If the DOE is correct, projections based on a logistic curve fit to the historical data cannot be used to make an accurate prediction. If Hubbert's model were correct, the coterminous US should only have about 20 billion barrels remaining, enough for about 13 years at the 2008 rate of production. Yet the DOE figure suggests over 20 times the amount that remains as recoverable in the entire US. Although the historical oil production values fit Hubbert's model, future production cannot conform to the declining leg of the predicted trend if the DOE is correct.

US production was more than replaced by the addition of new reserves during half the years between 1995 and 2008 (Figure 4.31). Given the

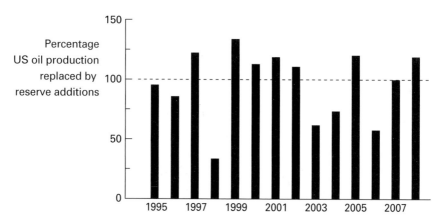

Figure 4.31   US replacements of oil production by additions to reserves from 1995 to 2008. Values include reserves and production from companies besides the large US-based corporations. About half the time, there has been addition to reserves from one year to the next. (Data: EIA)

year-to-year variations in production, a declining trend is difficult to detect. US oil reserves at the beginning of 2009 were slightly greater than what they were a decade earlier.

In 2007, major US oil companies replaced 105 percent of their US oil production by addition to reserves. Offshore reserve increases fell short of offshore production, and replacement was only 79 percent, but onshore replacement was 117 percent.[88] The large oil companies that report to the DOE showed an average US reserve replacement of 92 percent from 1981 to 2007. The replacement by large companies was only 65 percent since 2003, as they found more natural gas relative to oil. Even so, total US oil reserves declined by only 6 percent from 2003 to 2009.[89] Clearly, the US is a mature and heavily exploited oil region, and much production is in decline, but oil is still being discovered and significant production continues as the US remains the third largest oil producer in the world.

Although debate about diminishing US oil resources continues, in terms of the success rate of exploratory drilling, oil has become easier to find. Figure 4.32 shows the success of finding oil or gas in the US since 1950. In 1950, only one in five exploratory wells was successful. But the success rate from 2004 to 2007 ranged from 52 to 58 percent. In 1950, 2,010 successful oil and gas wells were drilled, while in 2007, 2,300 successful wells were drilled. However, in 1950, there were 10,300 exploratory drilling attempts to find petroleum, while in 2007, there were merely 4,400. The finds are smaller, but the likelihood of finding oil has increased. The majority of exploratory wells in the US are finding more natural gas than oil, just the reverse of the situation in 1950. In essence, the oil industry is better at finding oil and gas even though the exploration effort appears to be greater than it once was.

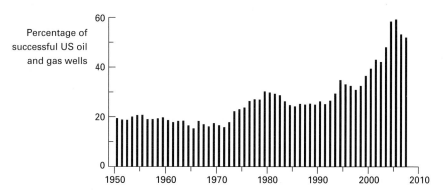

Figure 4.32 Percent of successful exploratory oil and natural gas wells since 1950. (Data: EIA)

Global Oil: Reserves

An even brighter reserve picture emerges worldwide. Globally, as the infla-tion-adjusted price of oil increased over the two past decades, reserves increased significantly. From 1986 through 2008, average annual oil prices (in 2007$) went from $32 to $94 per barrel, and during this period, global oil reserves grew from 0.7 to 1.3 trillion barrels. There is not a one-to-one cor-respondence between oil price and reserves, but the trend of increasing reserves with increasing oil price is suggested in Figure 4.33. After the early 1980s, inflated oil prices began to stabilize. From 1987 to 1990, both reserves and the price of oil increased by about 45 percent. During the 1990s, prices remained low, averaging less than $25 per barrel, and reserves remained flat. With the build-up to and entry into the US-led war in Iraq in 2003, oil prices more than doubled, and global reserves increased by about 25 percent. It is not likely a coincidence that with higher prices, more global oil resources have been converted into (profitable) reserves.

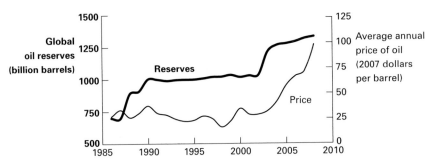

Figure 4.33   Oil reserves have increased with the increasing trend in oil price since the late 1990s. (Data: EIA)

Consider just the US and Canada. If oil in the US is considered together with that in Canada, new reserves have more than replaced current production in both countries. Canada's oil sands, through which liquid oil does not flow and which are therefore termed an unconventional oil source, are now a major component of the global oil landscape. Because of Canada's oil sands, as seen in Figure 4.34, its reserves jumped from 5 billion barrels in 2002 to 180 billion barrels in 2003, or from 2 percent to 69 percent of Saudi Arabia's massive conventional oil reserves. These oil sands were identified years ago, but they could not be profitably produced. In response to increased demand and price, however, the technology to profitably extract oil from oil sands was devel-oped. There were no surprise discoveries that led to Canada's increase in reserves but rather new economical oil recovery technology. It is important to note that the distinction between unconventional oil sands and conventional oil is artificial from an economic viewpoint. Both can be profitably extracted.

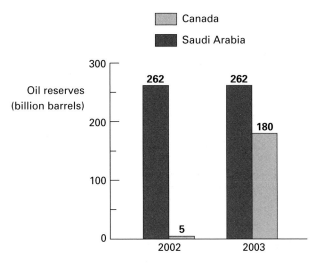

Figure 4.34   Reserve increase in Canada from 2002 to 2003 due to classification of oil sands as reserves. Canada is second to Saudi Arabia in terms of world oil reserves. (Data: EIA)

It is sometimes argued that the cost of finding and producing oil has increased as oil has become scarcer and more challenging to discover and exploit. That is not the obvious conclusion when one looks back at oil costs over the past 25 years. As shown in Figure 4.35, globally, total finding and production costs adjusted for inflation were less in 2007 than in the early 1980s. Costs declined through 2001 and rose with the rapidly increasing nominal price of oil from $25 in 2002 to $97 per barrel in 2008. Production costs have remained below about $10 per barrel (2007$) since 1985, increasing since 2005 but still below early 1980s values. Finding costs declined from over $17 per barrel in 1981 to about $15 per barrel in 2003. Since then, the cost of discovery climbed to $18 per barrel. Much of this worldwide increase is attributable to high US offshore finding costs, which rose to $65 per barrel in the 2004–6 period but declined to $50 per barrel in the 2005–7 period. These high US offshore finding costs are in stark contrast to much lower foreign offshore finding costs of $20 per barrel for the period 2004–7 (not shown in the figure).[90]

Global oil resources sufficient to meet demand have been maintained by a combination of new discoveries and reserve growth (more recoverable oil

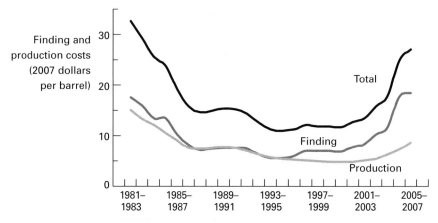

Figure 4.35   Global oil and gas finding and production costs since 1981. Three-year running average values for major US oil companies. Note that values are in 2007$ per barrel of oil equivalent. (Data: EIA)

than originally anticipated in existing fields). Consider global reserve growth. In its forecast, the USGS estimated that one-quarter of the global oil endowment (oil recoverable by 2025) would be derived from existing sources as reserve growth.

The USGS has analyzed field growth based on the US experience of historical reserve estimates for discoveries. In the US, reserve estimates for discoveries made since 1898 have been catalogued by the Department of Energy (Energy Information Administration). The USGS presented an analysis of this 100-year data set in 1998.[91] They found that over time more oil than expected was found in and produced from the previously discovered fields. Increasing US reserves did not require massive discoveries in unknown areas, but, rather, growth occurred as new oil in existing fields whose reserves were already catalogued.

The USGS developed a method, based on the US experience, to predict the likely addition to global reserves over 30 years (the time-frame of the USGS Assessment). The main factor used in their prediction was "oil field age since discovery." Age since discovery is a measure of technological improvements in equipment, methods, and oil-field production experience. The procedure followed by the USGS to estimate reserve growth was to retrieve field data and, for each oil field, predict the likely 30-year addition to reserves based on the discovery age of the field. Predictions were made for each field and compiled to create a global assessment of reserve growth. Figure 4.36 shows the field-growth function constructed from data on oil and natural gas fields in the US that was used by the USGS. To predict field growth, a known volume of oil in an existing field was multiplied by a 30-year reserve growth factor obtained from the function (in Figure 4.36).[92] For example, a five-year-old discovery is projected to double its reserves in 30 years. In its 2000 Assessment, the USGS reported the expected reserve growth as well as uncertainty bounds with optimistic and pessimistic values.

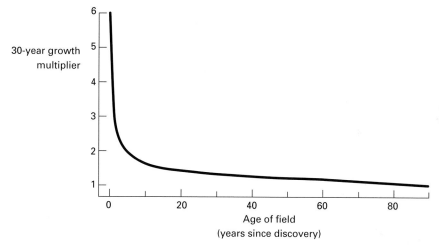

Figure 4.36   Thirty-year reserve growth versus years since discovery showing rapid growth in younger fields and stabilizing growth in older fields (after Schmoker and Klett, 2000[93]).

Is it reasonable to count on global reserve growth as a source of oil? Is the US reserve-growth experience applicable to the rest of the world? Given that the USGS forecast was based on a snapshot of the world's oil endowment taken in 1996, there have been many years of data to compare to its forecast. In 2005, the USGS published a post-audit of their 2000 Assessment, focusing on global reserve growth predictions since 1996 (snapshot date of the Assessment). Since the beginning of 1996, 2,142 petroleum accumulations had been discovered in known areas. In these areas, oil reserve growth for the period 1996 through 2003 was 28 percent of the previously assessed reserve estimate. This value is on target with the USGS estimate, given that 27 percent (eight years) of the 30-year time-frame of the Assessment had transpired. In other words, the USGS model of reserve growth, which was based on the analysis of historical reserve growth in the US, was on target. Natural gas field growth increased 51 percent during the eight years, suggesting that gas field reserve growth will dwarf the USGS estimates for 2025; that is, the

USGS underestimated natural gas reserve growth. During the post-audit period, reserve growth occurred primarily in the Middle East and northern Africa, as well as in Central and South America. The area of most new discoveries was Sub-Saharan Africa.[94]

The USGS post-audit showed that discoveries in new areas (versus reserve growth discussed above) of oil and gas comprised only about 10 percent of the reserve increase, which is short of the USGS projection, assuming a uniform discovery rate. However, three factors were not considered. First, the USGS post-audit was published in 2005, one year before the announcement of the huge new deep offshore discoveries beneath the Gulf of Mexico and off the coast of Brazil, so the USGS could not have included them in its report. Second, should oil prices resume their 2005–8 upward trend, it seems likely that the discovery rate will increase. Third, the USGS values had only considered conventional oil and gas resources and did not include the significant increase in Canadian oil-sand resources, which are now considered along with "conventional" oil and catalogued as part of worldwide reserves by *Oil and Gas Journal* and reported by the DOE's EIA. Had the USGS included all Canadian oil sands, oil discoveries would have reached 37 percent of the USGS total, again exceeding the interpolated gains of 27 percent. On the other hand, an extended period of low oil prices would discourage exploration, new offshore discoveries are increasingly finding natural gas versus oil, many prospects are in areas not open to exploration, and further discovery of unconventional oil resources may be hindered by the expense of oil recovery and environmental impacts.

Followers of Hubbert argue that reserve growth is accounted for in the Hubbert model by "backdating" reserve revisions to the original discovery date of each field. That is, each original oil field gets "credit" for the reserve no matter when the new oil is included in the reserve figure. However, by such backdating, the oil endowment value still increases with each revision of reserves. This strategy makes Hubbert's approach somewhat useless in that the value of the oil endowment is an ever-increasing number, as unforeseen reserve growth occurs. More importantly, a backdating system only accounts for past revisions and has no mechanism to account for future reserve growth other than to add more oil to the endowment, redo the analysis, and postpone the date of peak oil. This is exactly what some following Hubbert's approach have done.

Apart from forecast reserve growth, reported worldwide additions to oil reserves have surpassed global oil production over time. In fact, since 1990, reserves have increased enough to meet 170 percent of global production. From 1990 though 2008, cumulative oil production was 460 billion barrels. Additions to reserves not only matched that volume but grew by over 340 billion barrels. During this period, the life of global reserves increased by

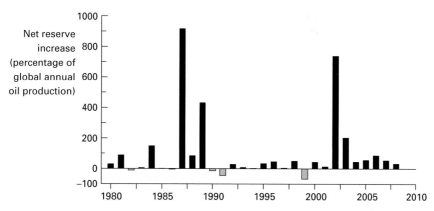

Figure 4.37   Percent of annual global oil production met by net reserve addition the following year. Values above zero indicate that there was a net gain in reserves after oil was produced. Values below zero occurred in years when net reserves declined from the previous year after oil was produced. (Data: EIA)

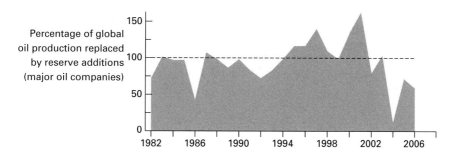

Figure 4.38   Global reserve replacement since 1981, as reported to the US DOE/EIA by major oil companies as part of the Financial Reporting System.[95]

over 12 years at the 2008 rate of production. Even if Canadian oil sands, which are now part of global reserves, are ignored, additions to "conventional reserves" exceeded cumulative global production by 165 billon barrels from 1990 through 2008. As seen in Figure 4.37, in most years, reserves have increased even after subtracting annual oil production. Some years were punctuated by large increases in reserves that wiped out historical deficits and maintained a net gain in reserves.

Records from the major oil companies that report to the US DOE show that global reserves exceeded 100 percent of production for the period from 1995 to 2001; that is, there was more than 100 percent reserve replacement (Figure 4.38). From 2002 to 2006, global reserve replacements dropped off to less than 59 percent, but is that declining trend likely to continue? Reserve replacement values have fluctuated. The quantity of global oil reserves (including NGLs) reported by major oil companies was fully replaced when summed over the period 1994 to 2006. In addition, natural gas reserves have been amply replaced by the major oil companies every year but one from 1994 to 2006.

Global Oil: Discoveries

What reason is there to believe that more oil is out there to be discovered? Globally, the number of new discoveries of giant oil and natural gas fields declined significantly after the 1970s. However, some modern trends are significant. First, the volume in giants discovered from 2000 through 2008 was higher than that discovered in the 1990s (57 versus 43 billion barrels). In the 1980s and 1990s, the average giant discovered had about 1.3 billion barrels of recoverable oil, while from 2000 through 2008, the average volume was 2.25 billion barrels. Of course, while these volumes are substantial, they are half the 4.5 billion barrel average of giants discovered in the 1940s through 1960s. The second trend is that the volume of condensate from natural gas is likely to be a significant source of oil. The volume of natural gas condensate relative to that of oil in global giant field discoveries has risen from less than 10 percent in the 1980s to over 50 percent in the 1990s, and to about 75 percent since 2000. For the period 2000 through 2008, the volume of oil as natural gas condensate in giant field discoveries (42 billion barrels) rivaled the volume of oil in new giant fields (57 billion barrels).[96] An increase in natural gas condensate is likely because about half of recent field discoveries have been natural gas rather than oil.

On average, natural gas giants discovered in the four decades since 1970 contain about twice as many barrels of oil equivalent as giant oil fields discovered during this period. The Sugar Loaf field, discovered in 2007 off the coast of Brazil, contains 33 billion barrels of recoverable natural gas condensate. Another discovery in 2008 off the coast of Brazil, the Jupiter field, is expected to produce 6.5 billion barrels of natural gas condensate.[97] Included in past discoveries are the South Pars gas and oil field in Iran, discovered in 1991 and ultimately producing 1.3 billion barrels of oil and 17.5 billion barrels of natural gas condensate;[98] the Kashagan field in Kazakhstan, discovered in 2000 and containing 11.6 to 13 billion barrels of oil equivalent, of which about 10 billion barrels are oil; and the Azadegan oil field in Iran, discovered in 1999 and containing 6.1 billion barrels of oil[99,100]. In mid-2007, China announced the discovery of the Bohai Bay oil field, containing 7.8

billion barrels.[101] Also discovered in 2007 was the Jubilee field beneath the waters of western Ghana, which may ultimately produce as much as 1.8 billion barrels.[102]

There are two insights that have been derived from discovery trends in recent years. First, more global offshore oil production is likely in the future. Data indicate that every 10,000 exploration wells consistently yield an addition of about 150 billion barrels of oil – about five years of global production at the 2008 rate.[103] Offshore exploration began in the 1940s. The result was 2,500 new discoveries on the back of almost 18,000 shallow-water (up to 1,500 feet of water) exploratory wells. Deep-water exploration (up to 7,000 feet of water) did not begin until almost 1980. Since then, 2,000 exploratory deep-water wells have been drilled with 400 new field discoveries. The net result of offshore exploration has been the discovery of 0.5 trillion barrels of oil, of which 41 percent has occurred in giant fields.

Major offshore discoveries and production are likely only if oil prices are high enough to justify exploration and development by companies and countries. From the 1960s through the 1980s, about one-third of giant fields were discovered offshore. The fraction rose in the 1990s and, since 2000, seven in ten new giant fields have been found offshore. Looking at all oil fields from 1999 to 2006, 140 fields were discovered globally; they contained 85 billion barrels; and they have the potential of producing upwards of 5.5 billion barrels per year. The majority, 127, of these finds were offshore, and 53 of them were in deep water (over 7,000 feet).[104] Global deep-water discoveries have accounted for 3 billion barrels per year, "replacing" about 10 percent of global oil consumption.[105] Offshore west Africa is an untapped resource: through 2005, there were 100 discoveries from 332 exploratory wells, and reserves are estimated at over 38 billion barrels.[106] Confirmation that large oil fields are still there for discovery, even in US offshore waters, occurred in late 2006, when Chevron announced a significant oil find beneath the Gulf of Mexico. At a depth of over 5 miles, this find contains anywhere between 3 and 15 billion barrels and could comprise 11 percent of US production by 2013.[107] In 2009, Chevron reported another deep-water discovery just 44 miles away that may yield 0.5 billion barrels and could be profitably produced at an oil price of $50 per barrel.[108]

The second insight from discovery trends is that an underlying premise of many statistical models of oil discovery is probably incorrect. This premise is that larger oil fields are found first, followed by the discovery of smaller fields. Large fields in geologically related proximity to one another are typically discovered first simply because they are the most easily detected targets. However, this is not always the case, as pointed out by Ron Charpentier of the USGS, who notes that new technology can rejuvenate the discovery process. That is, large fields in the same general region where oil was first found can be found by drilling deeper or in parts of the region where no oil was suspected to exist.[109]

Statistical extrapolation based on the notion that larger fields are discovered first, leaving only smaller fields behind, can seriously underestimate remaining oil resources. Two examples are the histories of discoveries in the Michigan Basin gas field in the US and the Trias/Ghadames Basin oil field in northern Africa. The Michigan Basin natural gas discoveries occurred in discrete periods of sequentially larger discoveries, starting in the 1940s, when **anticlinal traps** (subsurface mounds that can contain oil and gas) were easily detected, and continuing into the 1970s. Then improved seismic methods identified reservoirs contained in ancient coral reef deposits that trapped oil but in a geometry that was very different from the first-discovered anticlinal traps. In the case of northern Africa, large oil fields were first discovered in the 1960s, followed by a downward trend in new field sizes into the 1970s. However, fields just as large as the original 1960s' discoveries were made in the 1980s and 1990s based on improved exploration methods and enhanced seismic imaging of the geologic structures beneath large sand dunes.[110]

How well has the world been explored for giant fields? Discovery of oil in giant oil fields from 1956 to 1970 accounted for 525 billion barrels, or a rate of 35 billion barrels per year. During the subsequent 15-year period from 1971 to 1985, the discovery rate fell to 15 billion barrels per year (228 billion barrels total). The next period, 1986 through 2000, saw a further decline in the giant oil field discovery rate to 8 billion barrels per year (74 billion barrels total), and from 2001 through 2006, the rate dropped to 3.7 billion barrels per year. It was not until 2007 and 2008 that discoveries of giant oil fields picked up and the rate rebounded to over 11 billion barrels per year.

Why has there been such a dramatic decline in oil discoveries during the past 20 years? The first reason is obvious: depending on your definition of "easy," much of the "easy" oil has already been found. Finding oil and ultimately bringing it to market requires enormous investments. The second and more profound reason is that exploration was not vigorously pursued, or at least not pursued wisely by exploring in the most promising regions. With both low oil prices and meager profit margins from refining in the 1990s, investment in global oil exploration was depressed. This lack of vigor persisted for almost two decades, until recently. During the early 1980s, international oil companies reporting to the DOE spent about $35 billion per year on exploration. But from 1986 through 2004, investment in exploration dropped to about $10 billion per year (2007$).[111] Global expenditures in 2005 on exploration were less than half that in 1981. With the higher price of oil and greater exploration, the number of global discoveries of giant oil fields climbed from 1 in 2006 to 11 in 2008.[112]

As shown in Figure 4.39, the lull in exploration expenditures appears to be reversing course as oil companies allocate a greater portion of their resources to finding oil. Putting exploration investment in perspective, Chevron bought back $5 billion in stock in 2006, with exploration and capital spending of $17 billion and profits of $17 billion. Chevron increased its exploration and capital budget to $20 billion in 2007 and to $23 billion in both 2008 and 2009.[113] The giant Exxon Mobil spent $20 billion per year on capital investment and exploration in 2006 and 2007, which is less than the nearly $30 billion it spent on annual stock buy-backs. Yet, the company managed back-to-back $40 billion annual profits. By 2009, Exxon Mobil slowed its buy-back program in favor of expanding its search for new oil and gas. It plans to spend $150 billion through 2014 on drilling rigs, platforms, and refineries, with $29 billion in capital expenses for nine new projects in 2009. The company expects its oil and gas production to increase by 2 to 3 percent per year through 2014.[114]

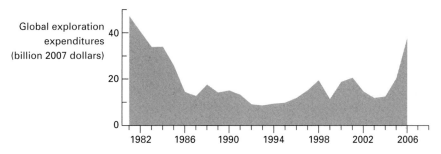

Figure 4.39 Global exploration expenditures by large oil companies reporting to the US DOE, showing reduced spending compared with 1981. (Data: EIA Financial Reporting System, 2007 – converted to 2007$)

Exploration and new production activity, as measured by the number of drilling rigs in operation, dropped precipitously after 1985 and only began to rise consistently during the past few years, as shown in Figure 4.40, which presents the Baker Hughes drill-rig count, an industry measure of the number of active drilling rigs. During 1986 to 2000, global drilling activity (averaging 1,893 rigs) was only 44 percent (1,893 versus 4,320 rigs) of that during 1980 to 1985, and the number in 2007, although up, was still just over half the peak level in 1981 (3,116 in 2007 versus 5,624 rigs in 1981). An interesting point, shown in Figure 4.41, is how limited worldwide drilling for exploration and production has been relative to that in the US, which currently accounts for over half of the worldwide total of active drill rigs. This percentage of drill rigs concentrated in the US has been fairly steady for the past 30 years, with a high of 71 percent of all worldwide drilling in 1981, when oil prices were near their maximum. In contrast, during the past 30 years, exploratory drilling in the Middle East, which contains the bulk of known oil, has averaged 10 percent of the worldwide total and reached a low of 3 percent in 1981. The oil industry was scouring the diminished oil reservoirs of the US, while not exploring in the Middle East even though there is likely much more oil to be found there.

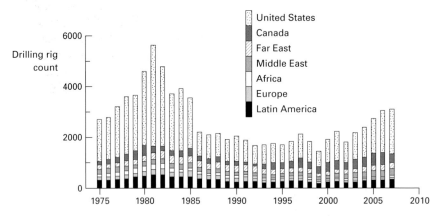

Figure 4.40   Drilling-rig count as a measure of exploration in different regions. (Data: Baker Hughes Inc.)

In essence, the US is a pincushion of exploration relative to other parts of the world, where more oil exists. As illustrated in Figure 4.41, the Middle East, Eastern Europe, and Africa contain 75 percent of world reserves and yet account for only 13 percent of exploratory drilling. Of the 15 giant oil fields discovered in the 1990s, four were in deep water off of Angola.[115] Even though west Africa accounted for one-fifth of the world's discovered oil reserves from 1996 to 2007, the number of exploration wells drilled there represented only 5 percent of the world total. The point is that meager exploration in much of the world has led to significant discoveries, and enhanced exploration in those regions is likely to find much more oil.

Just as remarkable, in 2005, the technical costs of oil – discovery, development, and production – in west Africa (Angola, Congo, and Equatorial Guinea) were only $8.70 per barrel, compared with the global average of $14.90 per barrel and $28.40 per barrel in the US.[116] The modern situation follows the pattern established in the past 30 years: the non-North American oil-rich regions that contain over three-quarters of global oil resources have an average of one-sixth of worldwide drilling (Figure 4.41).

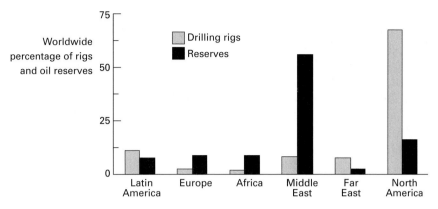

Figure 4.41 Percent of drilling rigs and oil reserves in different regions in 2008, showing misallocation of exploration resources.

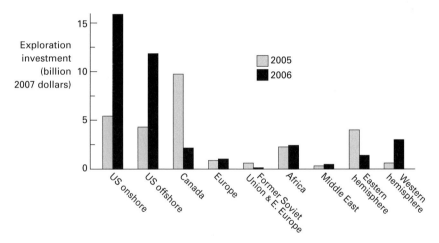

Figure 4.42 Global distribution of exploration investment by large oil companies in 2005 and 2006. (Data: EIA Financial Reporting System reports 2006 and 2007)

Reinforcing this argument, one can look at the global distribution of money invested in exploration in 2005 and 2006 by major oil companies (Figure 4.42). In 2005, fully 69 percent of investment was in North America, as compared to only 1 percent in the Middle East and 8 percent in Africa. Oil companies tripled their investments in US onshore and offshore exploration in 2006. The disparity in exploration effort is changing, as evidenced by investment data from 2000 to 2005. Even though heavily weighted toward the US, investment in exploration in Africa has more than doubled,[117] and exploration outlay in the Middle East has increased nearly sixfold.[118]

Oil-drilling exploration and new production have been mis-focused during the past three decades largely because areas with major potential, such as the former Soviet Union and the Middle East, have been off limits to private exploration. Arthur Berman, former Editor of the *Houston Geological Society Bulletin*, aptly notes that national oil companies control 80 percent of global oil reserves, and yet two-thirds of the discoveries made since 1999 have been made by just five private oil companies – British Petroleum, Exxon Mobil, Shell, Chevron, and Total.[119] Private investment in Brazil, Bolivia, Peru, Columbia, and Venezuela has resulted in significant increases in reserves during the past 10 years. As shown in Figure 4.43, South and Central American reserves have risen steadily by 30 percent during in the past decade. In Africa, reserves have increased by 69 percent over the same period. The above increments in reserves were largely due to private investment. By contrast, without private investment, Mexico's national oil company had a miserable rate of discovery, with its reserves declining from 47.8 to 10.5 billion barrels during the 10 years ending in January 2009.

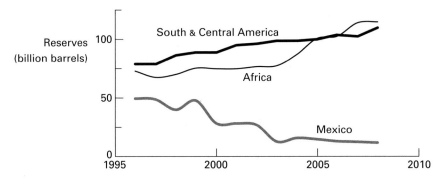

Figure 4.43   Private investment in Africa and South/Central America has resulted in an increase in reserves, while nationalization in Mexico has led to a decline in reserves. (Data: EIA)

Russian and Global Arctic Oil

If the world is not going to run out of oil, then there must be new sources. It is clear that much of the world's oil is concentrated in the Middle East, and the resources of that region have continued to be a guarded secret. But the Middle East is not the only promising prospect for global supply. Russian oil reserves of 60 billion barrels are being produced and accounted for over 13 percent of the 2008 world oil market (Figure 4.44), with production exceeding that of Saudi Arabia. Russian production has risen by over 50 percent since the mid-1990s, after the breakup of the Soviet Union and the reforming of the Russian economy. Russia's near-term economic future is tied to its energy resources. According to the World Bank, Russia's oil and gas might have accounted for up to one-quarter of its 2003 GDP, and each $10 per barrel rise in the price of oil bolsters Russian GDP by 3.5 percent.[120]

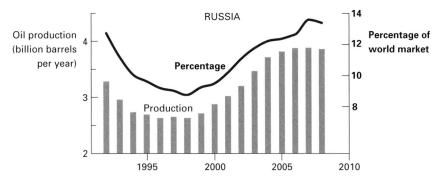

Figure 4.44   Resurgence of Russia as a major oil producer, capturing over 13 percent of the world market in 2007 and 2008. (Data: EIA and *Oil and Gas Journal* 2009)

Three-quarters of Russian oil has been produced from the Western Siberian Basin, a region that contains some of the richest petroleum accumulations on Earth. Spanning 2.2 million square kilometers (850,000 square miles), it is the largest petroleum basin in the world. About 150 million years ago, the central basin was occupied by a deep sea that covered more than 1 million square kilometers and in which sediments (forming shale) rich in organic matter were deposited on the oxygen-poor sea-bed. These rocks were responsible for generating 80 percent of the oil now found in the sandstones of the West Siberian Basin. Most of the giant oil and gas fields in the West Siberia Basin were discovered about 40 years ago, with production beginning in the early 1970s. The basin has been only moderately explored. Two potentially giant gas fields were discovered along Russia's coast in the Kara Sea, which is remarkable since only three exploratory wells were drilled.

How much oil is there in the West Siberian Basin? So far, 144 billion barrels have been discovered, and the estimate of undiscovered oil is 55 billion barrels.[121] Taken together, this amounts to three times the current Russian oil reserves, the eighth largest in the world in 2007. In addition, Russia has 1,300 trillion cubic feet of discovered natural gas, which is equivalent to 224 billion barrels of oil, and another estimated 643 trillion cubic feet is believed to exist (110 billion barrels of oil energy equivalent). Putting these numbers together, the oil and gas of West Siberia could amount to over 530 billion barrels of oil energy equivalent. This region alone could support about 20 years of current global energy-equivalent oil production.[122]

Although Russia has enormous oil and gas resources, it is unclear if it can continue to expand production in a timely fashion. The DOE reports about Russia that, "a 1998 USGS survey estimated that undiscovered, technically feasible, conventional reserves were larger than those of any other country in the world." A resurgence of Russian petroleum production followed the 1991 collapse of the Soviet Union, which in turn led to privatization in Russia, common use of Western technology, and better management of old oil fields. However, political and economic impediments exist for Russia to produce its plentiful oil.

In 2008, the USGS reported that, north of the Arctic Circle, there is potentially a lot of oil and even more natural gas. In all, this region is believed to contain 22 percent of the world's undiscovered, technically recoverable, conventional oil and gas, or over 400 billion barrels of oil and oil energy equivalent. The USGS appraisal of 25 geological provinces suggests that 90 billion barrels of oil is technically recoverable. Furthermore, 1,670 trillion cubic feet of natural gas (278 billion barrels of oil equivalent) and 44 billion barrels of natural gas liquids are estimated. The vast majority of the petroleum is offshore, with over one-third of the oil in Arctic Alaska and 40 percent of the gas in Western Siberia. Although oil and gas in some 400 fields has been found in this region, it remains essentially unexplored.[123]

Myth V: The World Cannot Afford Increases in Oil Use as Developing Nations Demand More Oil

When it comes to the affordability of oil and the demands of developing nations, there are actually two distinct fallacies that deserve discussion. The first has to do with the expected oil demands of developing nations and the second with oil as a controlling agent in the world economy.

Future Demand of Developing Nations

Perhaps the strongest argument supporting the concern about impending global oil depletion is that developing nations will need increasing quantities of oil as they industrialize. Worldwide oil demand will increase as the standard of living increases in emerging economies, such as those of China and India. Figure 4.45 (repeated here from Chapter 3) shows the relationship between oil consumption in barrels per person per year and economic well-being as measured by gross domestic product per person (GDP per capita) adjusted for inflation. As one might expect, developing nations have a very low intensity of oil use: that is, they use very little oil to run their economies. In stark contrast are the developed nations, which use much more oil per person to support their level of economic activity. The concern is that when China and India improve their standards of living, their annual oil consumption will not remain at 2.2 and 0.9 barrels per person but will mirror the US annual value of 25.1 barrels per person.[124]

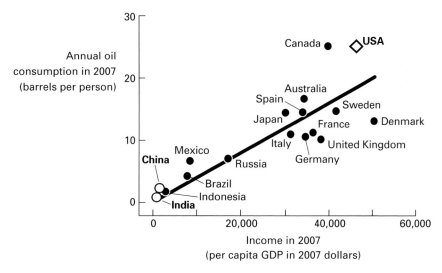

Figure 4.45 Oil consumption per capita versus income per capita in 2007 (2007$), based on gross domestic product (GDP) of countries shown. (Data: oil consumption, EIA; population and GDP, Economic Research Service, USDA)

Consider China. In terms of GDP, China's 2007 per capita income was one-twenty-sixth (4 percent) that of the US (in 2007, estimated average annual income in China was $1,970 versus $46,300 in the US (2007$)),[125] and its per capita oil use was less than one-tenth that of the US. Enormous demand for global oil resources would result if China's industrial development were more like that of the US in terms of industrial production, simultaneously increasing its per capita oil consumption and raising its standard of living. Suppose that China's population remained fixed at its 2008 value of 1.34 billion. Suppose further that China's annual oil use were to increase 11-fold to equal the US's 25.1 barrels per person, implying an increase in China's standard of living. Under this scenario, China would consume 33 billion barrels of oil per year and would exceed all annual global oil consumption (about 31 billion barrels in 2008)! Note that this comparison assumes zero population growth in China. If this becomes China's future, then a global oil depletion catastrophe indeed appears imminent.

What is not accounted for in the "when developing countries develop" scenario is that the patterns of consumption of developing nations have not been in lockstep with the historical patterns of developed nations like the US. In terms of oil use, the world has become much more energy- and oil-efficient during the past 25 years, particularly so in China. Globally, per capita oil production decreased by 17 percent from 1980 to 2007, and per capita consumption decreased by 10 percent, even though the **gross world product (GWP)** more than doubled in that 28-year period.

In some of the major industrialized, oil-consuming countries, such as the UK, Japan, and Switzerland, the *total* annual quantity of oil used (versus per capita use) has remained essentially unchanged compared with the early 1980s, and in France, Germany, Italy, Sweden, Finland, and Denmark, 2007 oil use (not in per capita terms) actually *declined* from values in 1980. In Eurasia, oil consumption is less than half of its value in 1980. In the Russian Federation (formed in December of 1991), oil use in 2007 was only two-thirds of its value in 1992.[126]

There is no doubt that the world as a whole is consuming more and more oil over time. However, how much more oil will be required depends substantially on two factors: (1) the worldwide efficiency of oil use, and (2) the rate of industrial development and growth of developing nations. Globally, **oil-use intensity**, defined as the oil consumption needed to produce national income on a per capita basis, has declined over the past 25 years. Consider the relationship of oil use and income back in 1980 (Figure 4.46) compared to 2007 (Figure 4.45). The trend line in 1980 was much steeper than that in 2007, indicating that it took much more oil per capita to produce the same amount of per capita income in 1980 than it did in 2007. Had one projected the trend in oil use in 1980 to 2007, assuming that the GDP of nations would increase, predicted oil use would have been greatly exaggerated (Figure 4.47). The decline in slope, which represents the reduction in oil-use intensity, is clear: the world has used relatively less oil to create greater income. On a worldwide basis, oil-use intensity declined by about 38 percent from 1980 to 2007. The global trend in oil-use intensity is displayed in Figure 4.48.

Over the 32 years from 1975 to 2007, oil use per capita in the US declined by 9 percent while, simultaneously, GDP per capita nearly doubled (in 2007$). Together, the reduction in oil use and increase in GDP have meant that US oil-use intensity dropped by about 54 percent over this period. The US used less than half as much oil per GDP in 2007 as it did in 1975 (Figure 4.49).

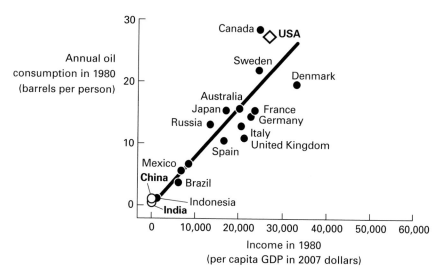

Figure 4.46 Oil use per capita versus per capita income in 1980 (in 2007$).[127] (Data: oil consumption, EIA; population and GDP, Economic Research Service, USDA)

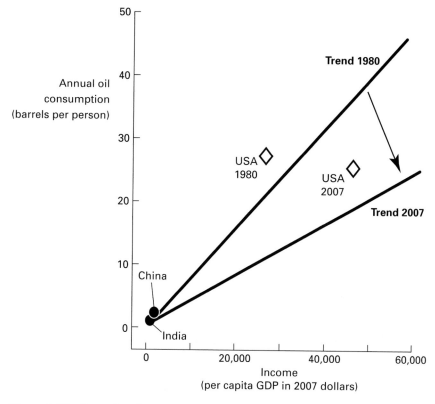

Figure 4.47 Per capita oil use versus per capita income in 2007 in comparison to the trend in 1980 (both in 2007$). Intensity of oil use (the slope of the trend line) has diminished since 1980. (Data: oil consumption, EIA; population and GDP, Economic Research Service, USDA)

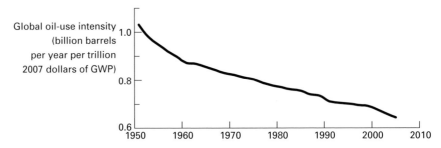

Figure 4.48   World oil use per GWP over time. (Data: oil consumption, EIA; gross world product and population, Economic Research Service, USDA)

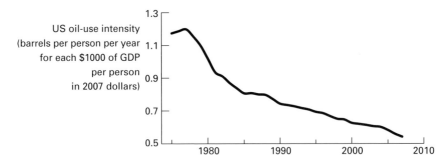

Figure 4.49   US oil use per GDP has declined for more than 30 years. (Data: oil consumption, EIA; GDP and population, Economic Research Service, USDA)

Oil-consumption data suggest that the world, and China in particular, has become even more efficient than the US. China has not adopted the historical oil-use pattern of modern-day industrial nations like the US. China's oil-use intensity declined by two-thirds between 1980 and 2007 – it uses one-third the oil it used in 1980 to generate one unit of income. The pattern of global oil consumption and industrialization suggests that developing nations likely will use less oil per capita to produce their incomes than that historically required by developed nations.

India's story is different from China's. Is India's economic growth a potential problem for global oil supplies? Perhaps, but not in the near term. Based on trends in per-person oil use per GDP in India, China, and the rest of the world, since 1980, India's oil-use intensity has indeed climbed, but it has remained low (Figure 4.50). India, with only half the GDP per capita of China, is extremely impoverished and has not rivaled China's oil consumption. India is early in the process of industrialization and has used 3 percent or less of the world's annual oil over the past 25 years. India has only recently started to become efficient in its use of oil to generate income. Since 2000, its oil-use intensity has declined significantly. Overall, India's pattern of oil consumption suggests that it will not impose a near-term stress on the world's oil supply, as it consumes oil at only about one-third the rate of China. However, like China, India has a population so large, over 1.1 billion, that even a modest increase in per capita use can pose a stress on global oil demand.

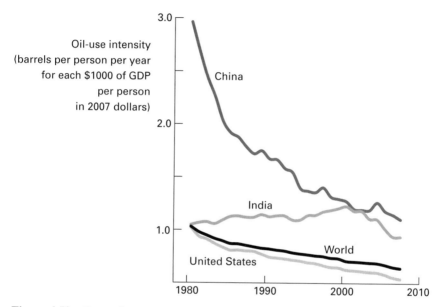

Figure 4.50   Comparison of oil-use intensities over time of the world, the US, China, and India. (Data: oil consumption, EIA; GDP, GWP, and population, Economic Research Service, USDA)

The general pattern of gain in efficiency spans the industrial nations; oil-use intensity has declined throughout developed countries in the past two decades (Figure 4.51). Although the world likely will require increasing amounts of oil during the next several decades, the oil requirements of developing nations are not likely to grow as rapidly as suggested by a forecast based on their historical consumption. There are two reasons for this. First, as discussed above, the oil requirements of developing nations have been offset by gains in the efficiency of oil use. That trend appears to be continuing. Second, several developing nations may use less energy as their economies move away from heavy industry. The US, which consumes 24 percent of the world's oil but has less than 5 percent of the world's population, is clearly a glutton nation when it comes to oil use. At the same time, the US accounts for 25 percent of the world's economy (21 percent based on purchasing power).[128] Should the world's industrial economy become increasingly dominated by developing nations, those nations are likely to need and use more energy, including oil, to produce goods for the rest of the world. If there is a transfer of industrial activity to nations like China and India, this may be accompanied by a relative decrease in demand on world energy resources by current industrial nations. For example, a reduction of US oil consumption by 12 percent would offset a doubling of India's oil demand. Even a modest decline in US oil consumption would make way for much of the needs of other nations.

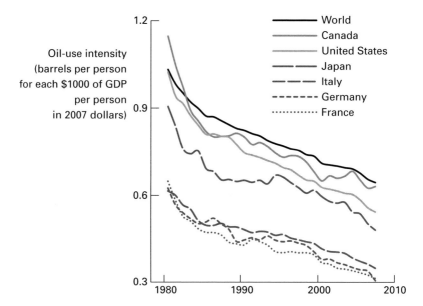

Figure 4.51   Oil-use intensity decline in developed nations since 1980. (Data: oil consumption, EIA; GDP and population, Economic Research Service, USDA)

Should global oil production peak and decline, it will likely reflect increasing demand for energy in developing nations. Through 2008, China's compound annual 10 percent economic growth was on the back of cheap energy and disregard for air, water, and land resources, and in some cases, consumer health. The response of the industrialized world is to wonder why China continues to choose prosperity at all expense. Why not learn from the mistakes of nations that have already followed this path? China's response is rhetorical – we are indeed following in your footsteps of industrialization and will worry about the environment later. Unfortunately, the economic development trajectory followed by the US and Europe over the past century may not work for China, which is compressing its industrialization into a period of mere decades. As the world has seen in the past few years, with higher oil demand comes higher transportation-energy prices. If the high energy prices of 2008 return, a consequence will be elevated costs of production and delivery of formerly inexpensive goods for export. If exports from newly industrialized regions become too expensive, global demand for the not-so-inexpensive imports will decrease. Sustained high oil prices would likely slow the growth of nations, such as China, on their fast track to economic development. High oil prices are incompatible with rapid economic development in industrializing nations if these nations plan to mimic the energy path followed by countries in North America and Western Europe. However, there would be counterbalancing feedbacks – slower industrial growth would temper oil demand and this would reduce upward pressure on oil prices.

Oil Expenditures in the World Economy

A major fear about the future is that the world cannot afford higher oil prices. This concern is based on the fallacy that the world is actually spending relatively more and more money on oil. In the 1970s, the OPEC embargo created a period of sustained, high oil prices that resulted in an economic recession (decline in real gross domestic product) and increase in inflation (rise in prices relative to purchasing power). Should oil prices return to their 2008 high levels and mimic the historical levels of the early 1980s, will the world economy be driven down further?

Viewed in historical context, global expenditures on oil in the past dozen years were not the highest ever. Rather, the 12-year period from 1974 to 1985 saw the most expensive oil in history. During that period, the world spent $15.4 trillion (2007$) on oil. In the more recent 12-year period (1997 to 2008), global expenditures were $13.7 trillion (2007$). In other words, the world spent more on oil during the last major oil crisis than during the period before 2009 when oil prices increased significantly.

What is even more significant is that oil has become a smaller cost-component of the global economy than it used to be. As a fraction of the average gross world product, over two times more money was spent on oil during 1974 to 1985 than during 1997 to 2008. During both of these periods, the price of oil increased rapidly. The average GWP was $21 trillion (2007$)

during the former period and doubled to an average of $42 trillion (2007$) during the latter one, and yet the world spent 12 percent less on oil from one period to the next (Figures 4.52 and 4.53). Another measure of the smaller role of oil in the world economy today is the relationship of global oil exports to world stock market capitalization. Although for somewhat different time periods, the *International Monetary Fund World Economic Outlook* notes that from 1973 to 1981, oil exports were 7.6 percent of stock market capitalization, while from 2002 to 2005, they were only 1.9 percent.[129]

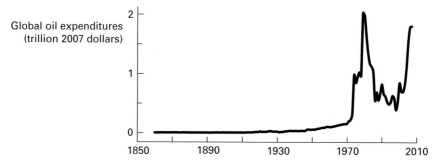

Figure 4.52   Annual global oil expenditures from first production to 2007, showing the costly period from 1974 to 1985. Note that estimates for 2008 and 2009 are not shown. Estimated 2008 global oil expenditure was $2.7 trillion. If the average oil price in 2009 were $50 per barrel, global oil expenditures would drop to approximately $1.4 trillion. (Data: oil price, BP; oil production, EIA)

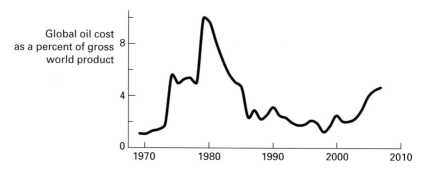

Figure 4.53   Cost of global oil expenditures relative to gross world product. In 1980, the cost of oil was twice the percentage of the world economy compared to the relative cost of oil in 2008. (Data: oil price, BP; GWP, Economic Research Service, USDA)

Recession was sparked in the 1970s when the (nominal) price of oil more than tripled in one year, going from $3.29 to $11.58 per barrel from 1973 to 1974. In inflation-adjusted terms, the price increased fivefold during the period 1973 to 1980 (from $15 to $93 per barrel (2007$)). Recessionary pressures diminished only when the price of oil dropped during the 1980s, falling by nearly half from 1985 to 1986 ($27.56 to $14.43 per barrel). Although the spot price of oil in 2008 exceeded $145 per barrel, the 2008 average oil price took five years to double in inflation-adjusted terms; the price rise was much slower than during the 1970s' oil shock. In terms of speed and magnitude, the oil price rise of 2008 itself was well tolerated by the economies of the world. It was the global financial crisis that began in 2008 that overwhelmed the world economy. Consequently demand and price dropped precipitously. Historically, the oil price increases of the 1860s, 1910s, and 1970s were each followed by a decade of declining oil prices.[130]

Before the economic crisis that began in 2008, the US economy stood up well to high oil prices for several reasons. First, just as oil is not as significant a component of the world economy as it once was, so too is oil not as major a component of the US economy. Oil expenditures in 1980 were 9.4 percent of US GDP, while in 2008 oil was "only" 5.0 percent of US GDP. As shown in Figure 4.54, in terms of consumers' pocketbooks, gasoline was a significantly greater portion of disposable income during the early 1980s (over 6 percent) than during the period since 2000 (less than 4 percent). A second reason for the resilience of the US economy in the face of the high oil prices of 2008 was that, although China consumed more and more of the world's oil, it produced cheap manufactured goods. Obtaining such inexpensive goods counterbalanced the inflationary impact of higher oil prices in the US. Third, a school of economic thought claims that there was a historical break in the effects of oil price shocks on economies such that subsequent oil shocks are not likely to be inflationary or cause a recession. Reasons for this break include decreased energy-use intensity, deregulation of key energy-producing and energy-consuming industries, and governmental monetary policy changes that have encouraged a low-inflation environment. However, as with many

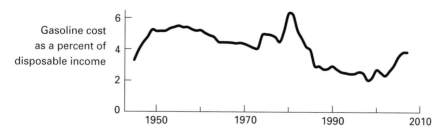

Figure 4.54   The cost of US gasoline consumption relative to personal disposable income. (Data: gasoline prices, EIA; US income, US Bureau of Economic Affairs)

economic theories, it is not conclusive that any of these links is responsible for the relative insensitivity of the US economy to oil shocks.[131,132]

Myth VI: There Are No Substitutes for Oil

Oil is different because there are no substitutes. This is another claim of those who maintain that the world is running out. But is it true? In terms of substitution, is oil different from other non-renewable commodities, such as copper and iron? Before answering this question, first we must consider what is meant by resource depletion. This is important because there is often little economic incentive to seek a substitute for a resource that is not scarce. To develop some intuition about the nature of resource depletion, we take a slight detour using the precious metal gold as an example.

The Gold Resource Pyramid

The globally available quantity of a particular commodity can be conceptualized as a "resource pyramid." The small volume near the top of the pyramid contains the reserves that are the easiest to access with simple technology, and the larger amount near the base represents the resources that are most technically challenging to extract. Take the familiar commodity gold.[133] Gold-rich source rocks, or "lode deposits" (as in "striking the mother lode"), occur in many places in the world, including South Africa, Western Australia, North and South America, and Siberia. Many of these deposits, which occur in veins and cracks in rock, were subject to millions of years of erosion. The heavy gold, in the form of nuggets, grains, and dust, was transported by rivers and subsequently concentrated in riverbeds as "placer deposits." Because it is much denser than typical river sediments, much of the gold has remained in riverbeds, where it is easily found and extracted by panning or other more elaborate yet similarly mechanical approaches to separate the dense gold from the lighter river sediments. Gold in placer deposits was easy to find and extract, in the sense that it took hard work but little technology to mine it. Placer gold discoveries spawned the California gold rush of 1849, the Australian gold rush of 1851, and the Yukon Klondike gold rush of 1897.[134] The easily mined gold in placer deposits can be represented as the region near the top of a resource pyramid, as shown in Figure 4.55. Of course, the placer deposits near the top of the pyramid were largely mined out over a century ago, so the pyramid represents "all" gold deposits that have been exploited to date as well as those that can be tapped in the future.

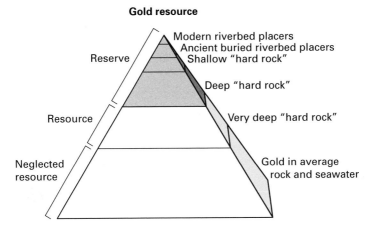

Figure 4.55   Gold resource pyramid showing reserves and resources, including gold already mined. The gold that can be extracted with simpler technology is initially the most accessible and lies at the top of the resource pyramid. The gold found in minute concentrations in typical rock and seawater is not counted as part of the gold resource.

Older sediments deposited by ancient rivers also contain gold, but the gold has been buried and initially was not easily mined. In California, shortly after riverbed mining began, sediments in cliff faces were mined using high-pressure water hoses to mechanically separate the gold from the gravel. The amount of ancient placer gold obtainable from "hydraulic mining" is represented as the next lower slice of the gold resource pyramid.

Once the easily found placer deposits were mined, the much larger quantity of gold extracted using the more advanced hydraulic mining technology made it cheaper and more profitable to mine gold compared with using the simpler approaches, such as panning. As such, moving down the resource pyramid reflects an advance in technology enabling greater quantities to be extracted and not necessarily a higher ultimate unit cost of extraction.

A still lower slice corresponds to gold that is deeply buried in rocks and therefore requires machinery for excavation and mechanical separation. For example, the Witwatersrand of South Africa contains massive amounts of gold that are mined to depths of 10,000 feet. Still lower down in the resource pyramid is the gold found in lower concentrations as small grains disseminated in large volumes of rock or associated with other mineral deposits, such as copper, that might not be economical to mine. Although the total quantity of "hard rock" deposits that could be mined is orders of magnitude larger than the gold found readily in riverbeds, recovery of metal from the rock requires arduous withdrawal followed by both physical and chemical processing. Many of these deposits are economically or technically feasible to mine and today represent the only significant source of gold. Other hard rock gold deposits, because they are too deep or too low grade ("dilute"), are not classified as reserves but rather as resources until price increases enough to make extraction profitable using existing technology.

Gold was selected as an example because it has a long history of exploitation and use, and relative to other metals there are few high-grade deposits. Globally, 193,000 metric tons have been mined. All of it would fit in a cube with sides of 21.5 metres (70.5 feet).[135] There are 100,000 metric tons (3.2 billion ounces) of gold resources, mostly in "hard rock," that have been identified (for which there is specific geologic evidence of the location, quantity, and quality) and about 18,000 metric tons that have been postulated and remain undiscovered. Under current conditions, less than half (47,000 metric tons) of all remaining gold is currently considered reserves.[136,137] Viewed from the perspective of remaining reserves, gold is relatively scarce.

There is a shortcoming in the above inventory of the global gold resource. If a complete gold inventory were wanted, we would include all the gold that exists in extremely low concentrations in many rocks, soil, and even dissolved in seawater. In Earth's crustal rocks, the average concentration of gold is about 4 parts per billion and in seawater, about 0.02 parts per trillion.[138] Based on average crustal abundance, rocks on Earth's dry land in just the upper 10 meters host over 13 million metric tons of gold, or 420 billion ounces.[139] This is 130 times the known gold on dry land that is considered available through mining. In addition, the oceans contain at least 26,000 metric tons of dissolved gold.

Although typically not even considered to be a part of gold resources, one could imagine counting the gold in average rock plus that in seawater and place this massive stock of gold at the base of the pyramid. However, extracting minute amounts of gold from typical rock or seawater requires advanced technology and is not economically feasible under current conditions. Gold in average rock and seawater is therefore also not part of reserves, but in principle it could be considered a resource because its existence is known even though it cannot be feasibly extracted at this time. But this gold is so dispersed and dilute that it is not considered part of the resource pyramid. When added up, this last form of gold represents the greatest quantity on Earth. However, it appears unlikely to be extracted even in the distant future. The point is that no resource assessment considers the most uneconomic, isolated, and dilute forms of a commodity, even if the total quantities are vast. In general, it does not make sense to include the most dispersed forms of a commodity in an economic resource assessment.

Two points about the gold resource pyramid are relevant to other commodities, such as oil. First, the volume of the pyramid represents the quantity of a resource and contains all forms of a commodity that have been and reasonably might be exploited for use. The cut-off for "reasonable" is subjective. Second, the upper portion of the pyramid corresponds to reserves (known, and technically recoverable under current economic conditions) plus all of the commodity that has been extracted to date, and the lower portion corresponds to resources (the likely additional ultimate source of supply). At any given time, a portion of the commodity exists as a reserve, and the remainder cannot be profitably extracted. This does not mean that resources lower down on the pyramid will always be much more expensive to extract than reserves higher in the pyramid. Rather, the lower portions of the pyramid represent the major fraction of a commodity that is more challenging to extract with technology and prices of the day. Technological advancements and price increases can rapidly improve the economics of extraction, converting large amounts of resources into (profitable) reserves. Consequently, the resource that is represented by the lower parts of the pyramid can be produced at a lower unit cost and can be sold at a lower price.

Where do we draw the line that distinguishes what to include as part of the minable gold inventory as represented in the resource pyramid? Take gold in seawater, for example. Although seemingly it is ridiculous to think that the ocean waters could be mined for gold, which is present in extremely low concentrations, imagine that gold was essential to human well-being and much more was needed than current supplies offer. Perhaps a type of genetically engineered microbe could be developed to concentrate gold in seawater over a period of years. Water-permeable but microbe-holding containers filled with these microbes might be deployed in the oceans at low cost, and mining the ocean's dissolved gold could be made economically feasible. Because of the vast quantity of gold in oceans, in principle, such a "fishing for gold" approach might be more cost-effective than obtaining gold from deposits on land, which requires expensive construction of mine shafts, excavation, extraction, and processing. Plus, these costs do not include the very negative impacts on the environment of conventional gold mining. The ocean gold resource could be converted into a reserve in the pyramid under the right technological and economic conditions. Nonetheless, mineral commodity analysts do not consider it likely that gold dissolved in seawater will be mined in the near future, so they do not include gold as part of the inventory of global gold resources.

Thus, the lower parts of the pyramid represent the fraction of a commodity that is more challenging to extract; but, ultimately, extraction becomes more profitable, and they become competitive with increasingly depleted, easily obtained reserves higher up the pyramid. That is, mining of a low-grade (low concentration) resource becomes price competitive with diminished reserves that provide the current supply. Initially, competitiveness might occur when prices reach a critical threshold, but as exploitation continues, technological improvements, cycles of technology innovation, and economies of scale create significant cost efficiencies.

The Oil Resource Pyramid

What about oil? Oil exploitation over time has sliced deeper and deeper into the conceptual oil resource pyramid. Conventional oil that flows as a liquid is at the top of the resource pyramid. When oil first began to be exploited, a well was drilled, and oil was either cheaply extracted or gushed out of the ground under natural pressure. In the very first developed oilfields, wells were drilled to very shallow depths, and liquid oil was easily exploited with technology of the day. For example, the first US oil well, drilled in 1859 by Edwin Drake in Pennsylvania, was just under 70 feet (21.3 meters) deep. Similarly, pioneering oil wells of the 1870s and 1880s drilled in the Soviet city of Baku (in modern-day Azerbaijan) were less than 250 feet deep, and the oil flowed under pressure to the surface. In 1901, Spindletop, Texas became the home of the gusher after oil trapped in salt domes and adjacent fault structures was hit at a depth of 1,020 feet. Today, conventional oil is obtained on land from wells drilled through miles of rock and offshore using platforms erected in thousands of feet of seawater. Yet, the price of oil averaged over the past several years is less than it was in 1864, when its inflation-adjusted annual price topped $106 per barrel (in 2007$). Throughout the period from 2000 to 2009, the inflation-adjusted price of oil was less than half that in 1864. The historical progression of new technology and cheaper means of extracting larger and larger quantities of global oil corresponds to slicing lower into the base of the oil resource pyramid.

More and more oil has been found by drilling deeper, but drilling depth is not the key point when considering the oil resource pyramid. Rather, with time we move lower on the pyramid as more challenging environments are targeted using novel technology that turns a dormant resource into a viable reserve. At first, new technology is very expensive, but costs decline over time, and the "old" new technology is eventually replaced by more advanced and initially more expensive "new" new technology. Rex Tillerson, Chair and CEO of Exxon Mobil, remarked, "… some today claim that the 'era of easy oil' is over. It is true that today oil is not 'easy' – but, over my 32 years, it has never been easy. ... Oil only seems easy after it has been discovered, developed and delivered."[140] Technological innovation reduces the cost of finding and developing a resource whose practical exploitation was almost unimaginable decades earlier.

The US and Global Oil Resource Pyramids

We can construct oil resource pyramids to represent estimated US and global oil supplies, respectively. Oil resources consist of conventional liquid oil and various forms of unconventional oil. Unconventional oil resources will be discussed in greater detail after the big picture is presented.

Consider the US oil resource pyramid (Figure 4.56). After production of 197 billion barrels in the US from 1859 to 2008, 21 billion barrels of proved US reserves remained at the beginning of 2009. It is these reserves plus the amount produced so far that sit atop the US oil resource pyramid. Conventional US oil has been acquired by slicing deeper through the pyramid with time: onshore oil deposits were extensively exploited first, followed by increased offshore oil exploration and production. Reserve growth and new discoveries of conventional oil account for almost 10 times (190 billion barrels) current US reserves (21 billion barrels). Moving further down the pyramid, oil left behind in the ground after initial pumping – stranded oil – is an even larger ultimate source of US oil (200 billion barrels).

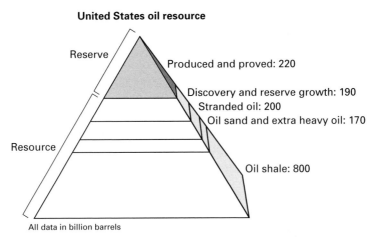

Figure 4.56 US oil resource schematic pyramid. (Data: DOE[142], EIA, and USGS)

Forming the base of the US pyramid is unconventional oil. The most technically accessible are heavy oils and oil sands, which together (170 billion barrels) are a source on par with all the oil ever produced in the coterminous US. At the base of the US oil resource pyramid is oil shale. This quantity of oil locked in US oil shale (800 billion barrels) is about four times the amount of conventional oil produced in the entire US to date and is equivalent to the conventional oil reserves of three Saudi Arabias.

To understand unconventional oil that makes up the lower portions of the oil resource pyramid, a brief description of the stages of oil extraction is needed. During **primary recovery**, oil is produced by taking advantage of natural pressure in the reservoir to force the oil to the surface through wells or by pumping the oil if the natural pressure is insufficient to extract the connected body of very fluid oil in place. **Secondary recovery** is used to improve production. This process entails gas injection or water flooding (injection) around the margins of the active oil field to sweep residual oil toward production wells. Finally, **tertiary recovery**, also known as **enhanced oil recovery (EOR)**, is used to obtain much of the remaining oil that is stubbornly stuck in place. Production of this residual oil is done by thinning the consistency of the residual oil so that it can flow to production wells by injecting gases, such as carbon dioxide, or heating the subsurface with hot water or steam.

Primary and secondary recovery are able to tap conventional oil that is represented by the top portion of the oil resource pyramid. Stranded oil is the next most challenging source and lies lower on the oil resource pyramid. Disconnected, large pockets of oil or dispersed residual oil trapped in isolated, tiny pores in the rock remain in the ground after the readily recoverable oil is extracted. As with gold, moving lower on the pyramid corresponds to that portion of the resource that is more dispersed, less concentrated, and/or more technically challenging to harvest. Also, as with gold, greater quantities of a resource are included as one moves lower down the pyramid. For example, in the US, there are an estimated 1.33 trillion barrels of oil originally in place. Of this total, the US DOE estimates that 390 billion barrels were or can be produced as conventional liquid oil, and that 200 billion of the 940 billion barrels that remain stranded can be produced using today's EOR methods. This value excludes production from oil sands.[141] In general, only about one-third of the amount originally in place can be extracted under natural pressure, pumped out (primary recovery), or forced out by water flooding (secondary recovery). Interestingly, although about 200 billion barrels have already been produced in the US, about 400 billion barrels remain technically recoverable. That is, about twice as much extractable oil as has ever been produced in the US remains in the ground!

Conservative estimates of global oil resources are shown in Figure 4.57. Since oil was first discovered, the world has produced about 1.1 trillion barrels through 2008. Reserves account for roughly 1.3 trillion barrels (1.15 not including Canadian oil sands). The USGS 2000 Assessment includes these reserves as part of their estimated 2 trillion barrels that comprise the remaining global endowment of conventional oil. The upper portion of the global oil resource pyramid represents the oil that has been produced through 2008. Lower in the pyramid are the reserves and the remaining technically recoverable oil. The base layers of the global oil resource pyramid signify the vast quantity of unconventional oil not included in the USGS 2000 Assessment but that exceeds all of the conventional oil that exists today plus all that has ever been withdrawn. The total conservative estimate of the global oil resource is about 6.7 trillion barrels, of which about one-sixth has been produced.

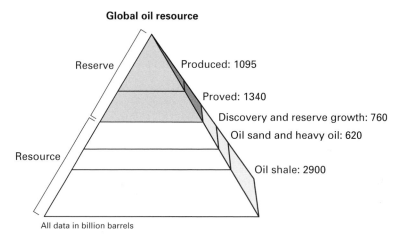

Figure 4.57 Conservative estimates of the global oil resource shown schematically as a pyramid. Proved reserves include Canadian oil sands (2009). (Data: USGS and EIA)

In 2005, the International Energy Agency (IEA, an organization with 28 member nations and not to be confused with the US DOE's EIA) estimated global availability of the conventional and unconventional oil sources shown in the oil resource pyramid and prices at which they become competitive, as displayed in Figure 4.58. As shown, greater and greater quantities of conventional (liquid) and unconventional (relatively or completely immobile) oil can be profitably recovered as the price increases. If the price of oil were above $80 per barrel (2007$), over 5.5 trillion barrels of oil can be profitably recovered (from conventional and unconventional sources), including the relatively small fraction of oil that has already been produced.

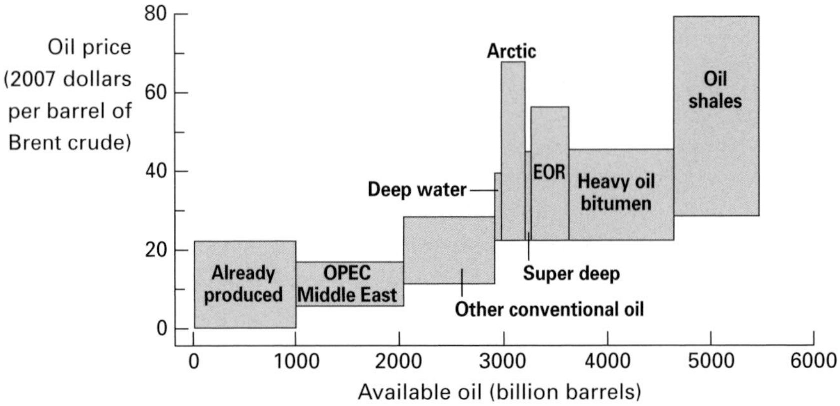

Figure 4.58   The relative price ranges in 2007$ for profitable extraction of the various forms of conventional and unconventional oil. EOR is "enhanced oil recovery." (*Resources and Reserves, Oil and Gas Technologies for the Energy Markets of the Future*, © OECD/IEA, Figure 7.1, p. 112, as modified by the author to reflect 2007$)[143]

Table 4.3   Estimated global initial in-place oil resources (trillion barrels)

	Middle East OPEC	Other OPEC	United States	Other non-OPEC	Global total beyond 2030
Conventional oil and condensate	2.6	2.6	0.9	2.9	9.0
Natural gas plant liquids	0.3	0.3	0.2	0.4	1.2
Heavy oil	0.0	2.3	0.0	0.0	2.3
Oil sands	0.0	0.0	0.0	2.4	2.4
Shale oil	0.0	0.0	2.1	0.7	2.8
Source rock	0.9	0.9	0.3	1.0	3.1
Total (corrected for rounding)	3.8	6.0	3.4	7.4	20.6

Data: EIA.[144]

In their long-term planning scenarios looking beyond the year 2030, the US Energy Information Administration employed estimates of initial in-place conventional and unconventional oil resources that suggest vast quantities remain worldwide; greater than the conservative estimates shown previously in Figure 4.57. The EIA values for the various categories of oil are displayed in Table 4.3. Only about 5 percent of the 20.6 trillion barrels estimated by the EIA have been produced to date. Just a portion (estimated to be less than 50 percent) of the EIA initial in-place oil is likely to be recovered, and the actual recovery fraction will depend on demand, price, and technology. The most significant difference between the in-place estimates of the EIA and the conservative estimates of recoverable oil presented previously is in the category of conventional oil and liquids produced from natural gas conden-sate. The global initial in-place total is 9 trillion barrels versus about 3.5 trillion based on the USGS estimates of oil and condensate that is technically recoverable by the year 2026. The EIA also assumes that 3.1 trillion barrels of oil ultimately could be extracted directly from oil source rock – a form of unconventional oil ignored in the USGS's conservative estimate.

Three Unconventional Oil Substitutes

Should it become so technologically challenging to extract a non-renewable resource that the market price increases and remains high, then substitutes for that commodity emerge. So what are the oil substitutes? Conventional oil is the liquid that historically has been pumped from subsurface reservoirs. The first set of substitutes for conventional oil are the various forms of unconventional oil. Unconventional oil is the term applied to oil that has not historically been pumped from the ground in large quantities because it does not readily flow under surface conditions. Most unconventional oil essentially sits immobile in the subsurface. Oil stranded in place, whose volume is about two times the oil originally pumped out from existing oil fields,[145] is the most accessible unconventional oil substitute. Beyond that, there are three other types of unconventional oil resources that exist naturally in a basically immo-bile form: heavy oil, oil sands, and oil shale. The following sections discuss each of these resources, first in the US and then globally.

US heavy oil

Heavy oil is much thicker than conventional oil and does not readily move towards an extraction well even when forced by pumping. Typically, to mobilize the oil, solvents are injected or heat is applied (via steam injection or in-situ combustion) (Figure 4.59). Today's technology has been used to economically produce heavy oil that lies up to 3,000 feet below the land surface. Excess costs over conventional extraction methods are $8 to $11 per barrel (2004$).[146] In the US, the current technically recoverable heavy oil lies less than 3,000 feet below the land surface. But somewhat more than half of US heavy oil deposits lie between 3,000 and 5,000 feet deep. Development of this resource, low on the US oil pyramid, requires advanced recovery technology, such as flooding with carbon dioxide gas at greater expense.

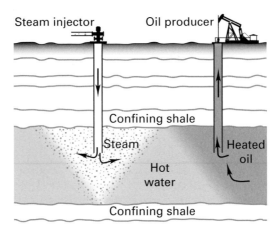

Figure 4.59   Steam flooding process. (The Kern County Oil Industry, copyright © 1998–2009 San Joaquin Geological Society, www.sjgs.com/steam.jpg. Reproduced in modified form with permission by the San Joaquin Geological Society)

There are 100 billion barrels of heavy oil in the US, more than half of which is in California and Alaska (42 and 25 billion barrels, respectively). In terms of the US oil resource pyramid, heavy oil adds a quantity that is almost half as much as the past production and current reserves combined. One major success in heavy oil recovery, which gives hope for this source of oil, has been the Kern River oil field in California. In 1942, it was estimated that 54 million barrels of heavy oil remained after primary extraction methods had been exhausted. By 1986, more than 13 times this amount, 736 million barrels, had been produced.[147] The DOE has estimated that with steam-based enhanced oil recovery technology, the expected yield from this one field is another 2.5 billion barrels, which is about two-thirds of the 3.9 billion barrels of the heavy oil that is expected to remain in place.

Global heavy oil

Heavy oil (including even thicker, extra-heavy oil) is found in exploitable deposits in over 30 countries. The USGS estimates that there are 434 billion barrels of technically recoverable heavy oil globally[148] (Figure 4.60). Not included in this estimate is additional heavy oil in the former Soviet Union (FSU), which has a reported 782 to 792 billion barrels, most of which is in the Siberian Platform. Of this FSU oil, approximately 19 percent is likely recoverable, yielding 150 billion barrels.[149] Counting the FSU deposits, the world has 584 billion barrels of recoverable heavy oil, or over 20 years' worth at the current rate of global oil production.

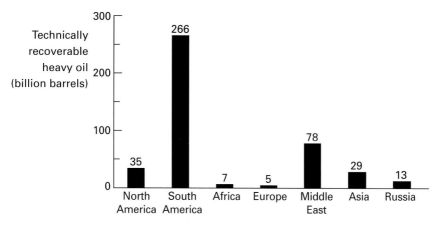

Figure 4.60 Technically recoverable heavy oil around the world. (Data: USGS, Meyer and Attanasi (2003)[150])

Of all technically recoverable heavy oil, 61 percent of the global total is in South America. Only 12 to 19 percent of the global heavy oil is technically recoverable, with current recovery rates often in the 5–10 percent range. The major reserves of heavy oil are in Venezuela, which is among the leading oil exporters in the world and a member of OPEC. Eastern Venezuela's Orinoco heavy-oil belt accounts for 90 percent of the global heavy oil resource, with recent estimates of Venezuelan oil ranging from 1.20 to 1.36 trillion barrels. The EIA reports the larger value of Venezuelan resources (1.36 trillion barrels of heavy oil).[151] The Orinoco belt holds 266 of the 584 billion barrels that can be technically recovered today. This heavy oil lies in a formation that is 600 to 3,600 feet deep. In the 1980s, commercial production of this oil began, with extraction through wells. The first petroleum to be produced was boiler fuel oil using a proprietary process, Orimulsion, which is a mixture of heavy oil, water, and 1 percent chemical surfactants (emulsifiers) to mobilize the oil. To make the heavy oil fluid enough to be transported, it must be diluted with substances such as natural gas liquids or light crude oil. In general, one barrel of other liquid is required to dilute and produce three to four barrels of heavy oil.

The rate of exploitation of heavy oil will depend on oil prices and the economic benefits of producing fuel oil versus other types of oil products. In 2003, Orinoco produced 156 million barrels. With the onset of high prices of oil in April of 2005, the Venezuelan national oil company (PDVSA) said that, beyond meeting its contractual obligations, it planned to stop production of its fuel oil and instead produce blended crude oil and synthetic crude (syn-crude).[152] It is important to bear in mind that although over 584 billion barrels of global heavy oil are technically recoverable, the assumed recovery is only about 15 percent (12 to 19 percent) of the trillions of barrels of the resource in place, which means that about 85 percent would remain in the ground, perhaps to be exploited in the future with new technology.

US oil sands

The second type of unconventional oil that is low on the US oil resource pyramid is oil sand, once called tar sand. Oil sand consists of immobile, semi-solid, heavy bitumen (a tar-like substance) in combination with water, sand, and clay. Bitumen comprises 1 to 20 percent of oil sands. When near the surface, the oil sands are recovered by mining. The bitumen is removed from the sand and clay, and then various processes, including coking, distillation, catalytic conversion, and hydrotreating, are used to make synthetic crude oil. On the other hand, when the oil sands are too deep to mine, the bitumen is mobilized by introducing steam or solvents, and then the bitumen is pumped from the ground. To produce a barrel of syncrude requires 1.16 barrels of bitumen.

US oil sands are estimated to contain 60 to 80 billion barrels of oil. Most of these oil sands are in Utah (19 to 32 billion barrels) and Alaska (19 billion barrels). There has been very little US production of oil sands for energy (less than 0.5 billion barrels cumulative production in California); some has been mined for road asphalt. US oil sands and heavy oil add 170 billion barrels to the lower slice of the US resource pyramid, not counting the deeper deposits for which new extraction technology will be required to make them economically viable.

Global oil sands

Of the global 220 billion barrels of oil sands, those in Alberta in western Canada have changed the global oil-reserve picture. With commercialization of this resource, Canada's reserves of about 180 billion barrels are now second only to Saudi Arabia's. Canada has 81 percent of the world's technically recoverable oil sands[153] and has been developing this resource with capital spending of US$8.25 billion in 2005 and annual production of 361 million barrels[154].

Most of the Canadian oil-sand deposits are within 2,000 feet of the surface, and part of the richest area, the Athabasca sands, is no deeper than 250 feet, so surface mining is practical in that area. The most recent estimate of in-place

Athabasca oil is 950 billion barrels. Canada's total oil-sands reserves have been estimated at 1.7 trillion barrels, although other estimates are higher (the USGS puts reserves at 2.7 trillion barrels, and the EIA at 2.37 trillion barrels).[155] Recovery is expected be 12 to 18 percent, and the best estimate of technically recoverable oil is 308 billion barrels, with about one-fifth coming from surface mining. Only 3 percent of reserves have been extracted since exploitation began in 1967.

Oil-sand recovery operations have employed both surface mining and in-situ extraction. Although surface mining is more common, the volume of oil from each process is expected to be about equal over the next 10 to 15 years. Surface mining requires physical separation of the bitumen and recovers almost 95 percent of the oil from the sand and clay.[156] Canadian Oil Sands Limited, Suncor Energy, and Shell Canada have all been engaged in oil-sand production. Suncor Energy, which began mining the Athabasca deposits in 1967, managed to lower production costs to US$10 per barrel in 2004. Suncor plans to nearly double production (to between 500,000 and 550,000 barrels per day) by 2012. Total surface mining in 2005 recovered 201 million barrels of Canadian oil from oil sands.[157]

In-situ production (steam injection and oil recovery) of Canadian oil sands totaled 160 million barrels in 2005. Forecasts by the Canadian Association of Petroleum Producers (Figure 4.61) show production of combined surface mining and in-situ production quintupling by 2025 to about 1.5 billion barrels per year. This is comparable to the current US oil production rate.

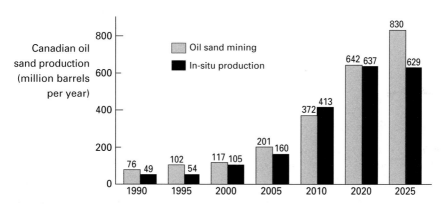

Figure 4.61   Projection of production of Canadian oil sands. (Data: Canadian Association of Petroleum Producers)

Surface mining of oil sands is profitable at a minimum oil price of US$22 per barrel and $30 per barrel if certain extra processing is required. In-situ oil sands can be profitably produced at US$38 per barrel. Below that price, about 30 percent of Canada's 2008 output would be halted, and at oil prices below US$30 per barrel, another 30 percent would fall below profitability. With the onset of the 2008 economic crisis and the rapid decline in oil prices, it was estimated that about $75 billion of new project investment would be halted, and the 2010 peak workforce needs would be cut in half from 44,000 to 22,000.[158]

US oil shale

Much lower on the US oil resource pyramid is the third type of unconventional oil that comes from **oil shale**. Oil shale is a hard rock, generally derived from calcium-rich clay and mud, that is loaded with kerogen. If this rock were buried under the right conditions (within what is called the oil window), it might produce oil naturally, over geologic time. Now tied up as a solid, the organic matter, kerogen, in oil shales needs to be artificially "cooked" over a human time scale to generate oil.

Today, oil shale is not being produced in the US, although Unocal did produce about 1 million barrels between 1957 and 1989. Like oil sands, oil from oil shale is obtained by two methods: surface "retorting" and in-situ recovery. In surface retorting, the mined oil shale is heated to 842 °F (450 °C) to yield gas and synthetic oil, mimicking in a compressed period the long geologic baking process that forms oil in Earth's crust. In-situ recovery involves heating the oil shale to generate oil and gas. In this process, 15 to 25 wells per acre are drilled through more than 1,000 feet of oil shale. The metal wells are heated to create temperatures of 650–700 °F (340–370 °C) in the rock over a period of two to three years. This process converts the kerogen to petroleum that consists of about two-thirds oil and one-third natural gas. The petroleum is then extracted by wells within the heated region (Figure 4.62). Shell, which developed the method, maintains that this approach is profitable if oil prices are in the mid-$20s per barrels. About half of the cost of this process is the energy needed for down-hole heating.

The fantastic promise of oil shale is the huge resource it represents. The Green River Formation in Colorado, Utah, and Wyoming is estimated to hold 1.5 to 1.8 trillion barrels of oil,[160] of which about 0.8 trillion barrels are potentially recoverable (Figure 4.63). Forming a huge base of the US oil resource pyramid, oil shales account for four times all past US production and reserves combined. Comparing Green River oil shales with the global endowment of conventional oil assessed by the USGS, they represent a resource equal to about one-quarter of the 3-trillion-barrel global oil

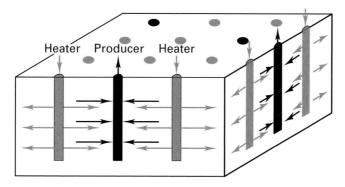

Figure 4.62 Proposed in-situ extraction technology for oil shale. (Adapted from Rand Report (2005)[159])

Figure 4.63 Location of the Green River oil shale. (Adapted from Rand Report (2005)[161] from material provided by Shell Exploration and Production Company)

endowment, or alternatively, they are triple the current reserves of Saudi Arabia. This oil shale would be sufficient for over 100 years at today's US rate of consumption.

As an energy source, the main drawback of oil shale is that it is a very low-energy-content fuel. It contains about one-eighth the energy of conventional oil on an energy-per-ton basis (5 million British thermal units (BTU) per ton versus almost 40 million BTU per ton in crude oil). It is not economic to exploit oil shales that would generate less than about 15 gallons of oil per ton. A 2005 study by the Rand Corporation concluded that oil prices would have to remain above $70 to $95 (2005$) per barrel for the first-generation mining and surface retorting plants to be profitable using existing technology. Such an oil price was surpassed in 2008, but it remains uncertain that such a price would be sustained in the future to justify commercial investment in large-scale recovery. Earlier Rand studies suggested that after 0.5 billion barrels were produced, the technology would improve, production costs would drop by 50 percent, and the oil shale resource would be profitable at $35 to $48 per barrel (2005$). The expectation is that such production efficiency would occur within 12 years of initial commercial operation. An analysis conducted in 2008 claims that US oil shale can be profitably produced at an oil price of $38 to $62 (2007$) per barrel.[162]

Global oil shale

With potential recovery of about 3 trillion barrels globally, there is about as much oil in oil shale as the total global endowment estimated in the USGS 2000 Assessment. However, within the oil industry, the standing joke is that, "Oil shale is the energy source of the future and always will be." Oil shale was mined from 1815 through the 1850s in both the US and Canada, until it was rendered uneconomic by the production of conventional oil in 1859. Such has been the history of significant commercial oil shale exploitation. Each time oil prices rise, there is renewed interest in development of this resource, but as prices have fallen, interest has waned. This may be happening again. For example, oil shale is used on a small scale as a coal substitute to run power plants in Israel, and obtaining oil from shale was considered competitive at $18 per barrel to the mid-$20s per barrel oil a decade ago, but large-scale production never materialized as oil prices fell. In July 2006, when oil prices rose above $70 per barrel, Israel again gave serious consideration to exploiting its oil shale to reduce its oil imports by 30 percent. Although the Israelis hope to use a new catalytic process to extract the shale oil, they are hampered by the especially low energy content of their oil shale. Even so, the process is purportedly competitive with oil at $17 (2006$) per barrel.[163]

If such new technology were to take hold throughout the world, oil shale could become a critical source of oil.

Oil shale is so far down on the global oil resource pyramid that there are no tight estimates of the total resource. There are documented reserves in China, Brazil, and Estonia, where oil shale has been mined and used, and other countries contain substantial oil shale deposits (e.g., Australia, Russia, Italy, Morocco, and Zaire). Worldwide, a credible published estimate by the USGS is 2.9 trillion barrels of recoverable oil, based on analysis of oil shale in 33 countries that contain a total of 411 billion tons of oil shale, which "should be considered a minimum figure because numerous deposits are still largely unexplored or were not included in [their] study."[164] The average of published, but uncertain, estimates of all global oil shale resources suggests that the quantity of oil shale containing at least 10 gallons of oil per ton is staggering: about 330 trillion barrels, which, even at 1 percent recovery, would roughly equal the current estimate of the global endowment of conventional oil (approximately 3 trillion barrels).[165]

Fossil Fuel Conversion: The Role of Gas and Coal

If the supply of conventional oil were to fall short, and the technology to economically extract unconventional oil from as yet untapped resources like oil shale did not materialize, what could possibly replace common gasoline, jet fuel, and diesel fuel, the most precious products derived from 60 percent of each barrel of oil? In fact, there are two fossil fuels besides oil that can be converted into liquid fuels: natural gas and coal. Abundant supplies of each exist, and the technology to convert either of these fuel stocks into liquid fuel has been viable for 70 years. Development of these resources would reduce the demand for crude oil.

By the time World War I ended in 1918, Germany had become reliant on liquid fuels to power its automobiles, ships, and industry. The difficulty was that Germany was an oil-poor nation. However, it was rich in coal resources, and so it embarked on an effort to convert its abundant coal into liquid fuel. Success was not long in coming. In the 1920s, Franz Fischer and Hans Tropsch, working at the Kaiser-Wilhelm Institute for Coal Research, developed the technology to synthesize liquid fuel from coal. The **Fischer-Tropsch process** may stand as one of the most important practical scientific breakthroughs in history, even if its commercial viability was entangled in Germany's war efforts both before and during World War II.

During the decade following its discovery, the Fischer-Tropsch process was improved to yield higher quantities of motor fuel relative to lubricants and waxes. By 1935, there were four German Fischer-Tropsch plants producing 800,000 barrels of liquid hydrocarbons per year, 72 percent of which was motor fuel. In preparation for war by 1940, Hitler announced his Four Year Plan at the Nazi party rally in Nürnberg in 1936. He put Hermann Göring in charge of the plan and made synthetic fuel development its central feature. Over 40 percent of the projects funded by the plan in 1936 and 1937 were aimed at generating fuels from Germany's natural coal and tar. Synthetic fuel enabled Germany's war effort. By 1944, production from nine plants reached 4.1 million barrels per year (330,000 barrels per month), but Allied bombing during World War II reduced that figure to less than 30,000 barrels per month by March of 1945.[166]

The Fischer-Tropsch process has been an intensive area of research, with over 1,000 related patents issued in the US since the 1920s and over 250 since 1991. Fischer-Tropsch is a gas-to-liquid (**GTL**) process. The basic idea is that a fuel stock, in this case natural gas or coal, is converted into carbon monoxide and hydrogen, which is called syngas (synthetic gas). This conversion can be done through coal gasification by heating and partial combustion of coal (so carbon monoxide rather than carbon dioxide is produced), or, in the case of natural gas, combining it with oxygen or alternatively adding water as steam (steam reforming).[167] The Fischer-Tropsch process then takes the syngas and uses catalysts (substances that promote but are not consumed by a reaction) to create a suite of hydrocarbon liquids. The catalysts are commonly cobalt, when the feedstock is natural gas, and iron-based substances, when coal is the hydrocarbon source. Depending on the temperature of the reaction, different suites of liquids are produced. For example, the high-temperature Fischer-Tropsch process yields lighter liquids containing about 36 percent gasoline, while the low-temperature process yields heavier fractions and only about half as much gasoline. Some of the longer-chain hydrocarbons (heavier fractions) are "cracked" to produce diesel by post-treatment with hydrogen. Because impurities are removed during different stages of the gas-to-liquid process, the diesel fuel that results is low in sulfur and produces fewer particulates when combusted than diesel traditionally refined from oil.[168,169] One potential breakthrough in the process came in 2006, when Rutgers University professor Alan Goldman and University of North Carolina professor Maurice Brookhart discovered a new pair of catalysts that work in tandem and, when made a part of a Fischer-Tropsch process, can yield a much higher fraction of diesel fuel than was possible with single catalysts. The method employing this exotic catalyst pair – one iridium-based and the other molybdenum-based – has not been commercialized.[170]

In general, the gas-to-liquid process yields a type of petroleum with three main categories of products: fuels, lubricants and waxes, and petrochemicals (naptha). Of key interest in terms of substitution for oil is the production of diesel fuel, which today represents 50 to 80 percent of GTL products. Why is diesel fuel so important to the future of oil?

The Importance of Diesel

In the US, diesel has a bad reputation as a dirty-burning fuel that is often difficult to find at gasoline stations. But the fact remains that Rudolf Diesel's 1892 engine has always been far more efficient than traditional gasoline combustion engines because of its high compression ratio, and the diesel engine is potentially cleaner-burning than traditional combustion engines. In terms of miles per gallon, diesel engines are about 30 percent more efficient than gasoline engines. In terms of emissions, the US has historically produced a type of diesel fuel that has relatively high emissions and poor ignition quality, but, since 2006, more efficiently burning, highly refined, ultra-low sulfur diesel has become more widely available, and traditional US diesel fuel is scheduled to be eliminated by 2010. Although the number of registered passenger diesel vehicles in the US grew by 43 percent from 2000 to 2005, they only represented 3.6 percent of the total number of registered passenger vehicles.[171] Use of diesel fuel in the US has increased by 5 percent per year over the past decade, but diesel is not commonly available at filling stations. In Europe, where diesel cars are common, diesel is a popular automotive fuel available at most filling stations, and the already high standards set for diesel emissions are improving dramatically.[172] Sales of diesel-fueled cars have grown in Europe and represented 50 percent of car sales in 2007 (Figure 4.64).[173] In 2008, estimated diesel vehicle sales constituted 60 percent of the market in France and Austria. Diesel vehicles represent 20 percent of the fleet in Japan.[174] Globally, the use of diesel fuel is growing steadily at about 3 percent per year.

Synthetic Fuel from Coal and Natural Gas

Synthetic fuel has been tested in the US with good results. In a one-month trial in 2002, synthetic fuel made from natural gas was used without performance or maintenance problems in heavy-duty vehicles. In 2004, a 12-month trial showed a significant reduction in emissions compared to typical diesel fuel. Compared with diesel sold in California, where these trials took place, synfuel produced 22 percent fewer exhaust hydrocarbons, 38 percent less carbon monoxide, and 30 percent fewer particulates.[175]

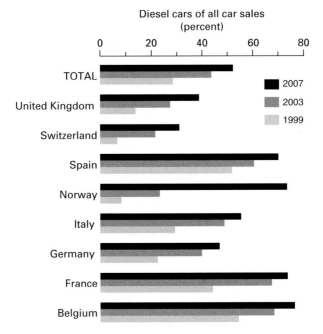

Figure 4.64   Increase in sales of diesel cars in Europe. Total includes Austria, Belgium, Denmark, Ireland, Finland, France, Germany, Greece, Iceland, Italy, Luxembourg, the Netherlands, Norway, Portugal, Spain, Sweden, Switzerland, and the United Kingdom. (Data: *Automotive Industry Data Newsletter*, Nos. 0102, 0302, 0501, and 0602)

The US Department of Defense tested a 50:50 mixture of traditional jet fuel and synfuel produced from natural gas using the Fischer-Tropsch process. It found that the mixture resulted in 50 to 90 percent fewer particulates and no sulfur. In 2006, the US Air Force spent $10 million per day on fuel, so, for example, a $15-per-barrel oil price increase translated into $900 million more expenditure per year. Their hope is to have a reliable and affordable supply based at least in part on synfuel.[176]

Commercial plants currently produce transportation fuels from coal and gas using the Fischer-Tropsch process. The company Sasol in South Africa operates three plants with a total capacity of 30 million barrels per year using coal as a fuel stock; these plants have produced over 1.5 billion barrels of oil equivalent fuel. In Qatar, Sasol is responsible for building a GTL plant that will produce 12 million barrels per year.[177] In the US, three small, coal-based plants are being developed in Pennsylvania, Iowa, and Colorado. Montana has plans for a plant that will produce 8 million barrels per year.[178] China, which is rich in coal, is developing plans to construct two coal-to-liquid plants, each capable of producing 29 million barrels per year.[179] Shenhua China Coal Liquefaction Corporation is constructing a 7 million barrel per year plant in Inner Mongolia using a direct liquefaction approach. In Shaanxi Province, a 14 million barrel per year plant is under construction by Yankuang Coal Mining.[180–182] China's long-term plan is to build more plants to produce a total of 1 million barrels per day by 2020.[183] However, recent environmental and cost concerns could delay China's plans for liquid fuel obtained from coal.[184]

Other countries could benefit from gas-to-liquid technology, especially those that discard their natural gas as an unwanted by-product of oil production, as sometimes still occurs. For example, in 2000 through 2004 in Africa, about 1,300 billion cubic feet of natural gas was flared, which is equivalent in energy to 217 million barrels of oil per year.[185] Worldwide, over two times this amount (2,741 billion cubic feet) was flared in 2004, which was only 2 to 3 percent of global production, but still represents about 450 million barrels of oil equivalent.[186]

Natural Gas Resources

Natural gas can serve as a substitute for oil used for transportation for cars and trucks equipped to run on it. Natural gas can also be converted into diesel fuel using gas-to-liquid processing. The USGS 2000 Assessment estimated global natural gas reserves at 15.4 quadrillion cubic feet, which is equivalent in energy content to 2.6 trillion barrels of oil.[187] This value is nearly on par with the global oil endowment (3 trillion barrels), except that comparatively little natural gas has been consumed. In 1996, the consumption of natural gas accounted for only 11 percent of the gas endowment (Figure 4.65). Since that time, cumulative natural gas consumption has increased, and by 2008 the value was closer to 17 percent – still a small fraction of the total endowment. At the 2008 rate of production, the estimated natural gas endowment would not be exhausted for 125 years. Not only is there a rich endowment of natural gas, but reserves are substantial and have grown over time: the ratio of reserves to production has risen to over 70, having climbed from 40 in 1960. In both absolute terms and relative to consumption, there are more (known) gas reserves today than in the past.

Unlike oil, for which the Middle East accounts for 60 percent of reserves and 45 percent of the endowment, the Middle East "only" accounts for 40 percent of natural gas reserves and less than 30 percent of the natural gas endowment. About one-third of the global natural gas endowment is in the

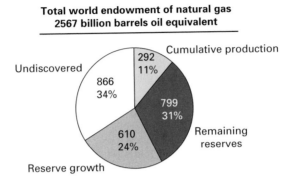

**Total world endowment of natural gas
2567 billion barrels oil equivalent**

Figure 4.65   Global endowment of natural gas based on the USGS 2000 Assessment. The assessment covered the period 1996–2025, so the figure shows the breakdown of the segments as of 1996. (Data: USGS)

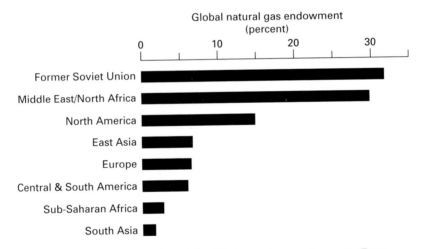

Figure 4.66   Regional distribution of the global natural gas endowment. (Data: USGS)

former Soviet Union, and 15 percent is in North America, of which over 80 percent is in the US (Figure 4.66).

Combining natural gas and oil endowment figures, it is evident that the world had consumed only one-fifth of the total petroleum resource endowment by 1996 (the Assessment start date), and remaining reserves, reserve growth, or undiscovered quantities represent substantial future supplies (Figure 4.67). It is important to remember that the USGS Assessment did not include what has historically been called unconventional oil, nor did it evaluate hundreds of geologic provinces where oil might exist. If natural gas liquids were considered, the global petroleum resource endowment would increase by 324 billion barrels.

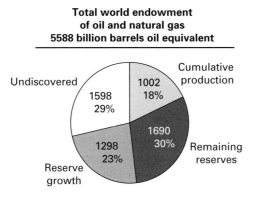

Figure 4.67 Total petroleum endowment breakdown based on the USGS 2000 Assessment (given the Assessment start date of 1996). (Data: USGS)

The USGS conducted a "reality check" of the natural gas endowment figures published in the 2000 Assessment, just as it did for its global oil endowment estimate. Recalling that the cut-off date for data used in the USGS 2000 Assessment was 1996, the USGS conducted a post-audit through 2003 of its 30-year projections for natural gas. The USGS found that within the eight-year period since its Assessment, discoveries had increased by only about 10 percent, but reserves had grown significantly – by 51 percent. On balance, the total of discoveries and reserve growth of natural gas was 27 percent, which corresponds well with expectations based on the USGS 2000 Assessment, considering the post-audit period of eight years is 27 percent of its 30-year assessment time-frame.[188] In other words, 27 percent of projected new natural gas reserves were added during 27 percent of the 30-year horizon of the USGS Assessment. Interestingly, the USGS might have greatly under-estimated the quantity of natural gas in its 2000 Assessment. As the authors of the Assessment review noted:

> The USGS assessment is not exhaustive, because it does not cover all sedimentary basins of the world. Relatively small volumes of oil or gas have been found in an additional 279 provinces, and significant accumulations may occur in these or other basins that were not assessed. The estimates are therefore conservative."[189]

Just as in the case of sequential assessments of the global oil endowment, estimates of ultimately recoverable natural gas have climbed since the first estimates were made. Initial estimates in the late 1950s and early 1960s, including one by Hubbert in 1962, were about half of the USGS 2000 Assessment value. In the 1970s and 1980s, 28 different published estimates averaged about 66 percent of the Assessment value, and in the 1990s, 13 different published estimates averaged about 80 percent of the USGS value.

Coal Resources

The remains of plants that have been buried, compacted, and aged several tens of millions to hundreds of millions of years, coal forms by a progressive removal of the water contained in the source material and a simultaneous increase in the fraction of carbon and stored energy. Coal's quality ranges from lower-energy-content lignite, to higher-energy bituminous coal, to, ultimately, anthracite. Coal has been used as a fuel for at least 2,000 years, with coal mining beginning by the thirteenth century in Europe and continuing to this day. By 1900, coal supplied 90 to 95 percent of the world's commercial energy. As oil became more widely used, this percentage declined to less than half by 1955 and stands at about one-quarter today. Although coal has declined relatively as a global energy source, because of its tremendous global abundance, it has significant potential to provide liquid fuels for transportation and thereby substitute for oil's most significant use.

Estimates of the global coal resource by both the DOE and British Petroleum converge at about 11 trillion metric tons. At even three times the current rate of production, the global coal endowment is sufficient to last over 500 years. There are almost a trillion metric tons[190] (0.85 tmt) of proved recoverable coal reserves, which, at current rates of production, would last 156 to 164 years.[191,192] The US and the former Soviet Union contain more than half of the known coal reserves (Figure 4.68). These coal reserves are equivalent to 6.7 trillion barrels of oil, or more than twice the global oil endowment estimated by the USGS in its 2000 Assessment. Just considering the US, the coal reserve-to-production ratio is about 240; that is, coal would last 240 years assuming today's reserve estimates and production.

The global distribution of coal resources (versus reserves) is less restricted than oil resources, and the major coal source areas do not coincide with those of oil. Over 80 percent of the coal lies in four regions: the former Soviet Union contains about 50 percent, the US 15 percent, China 10 percent, and Australia 7.5 percent. Most of the remaining substantial coal deposits lie in India, the UK, Germany, Poland, and South Africa. Should coal-to-liquid

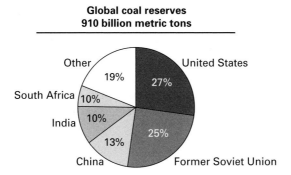

Figure 4.68 Global coal reserves in 2007. Globally, 53 percent is anthracite and bituminous coal, 30 percent sub-bituminous coal, and 17 percent lignite. (Data: EIA)

technology be adopted more widely in the future, there is certainly ample coal in currently stable parts of the world to serve as supplies. However, carbon emissions from the use of coal is a major concern. Based on the efficiency of diesel engines and the abundance of fuel stocks (both natural gas and coal) for conversion to diesel fuel, in the near future diesel could displace gasoline as the preferred fuel for automobiles that run on liquid fuel.

Chapter Summary

- All of Hubbert's forecasts of today's coterminous US oil production are much lower than the actual values.
- The trends in production of global oil and natural gas have not declined as predicted.
- The world has never run out of any significant globally traded, non-renewable Earth resource.
- Peak production and decline have occurred for various renewable and non-renewable resources due to decreases in demand. Substitution to meet end-use needs has occurred for a variety of non-renewable resources.
- Global oil production has not followed the pre-1980 trend. The increase in global oil production has tracked directly with the increase in global population. For the past 25 years, global per capita oil production has been relatively steady at 4.1 barrels per person per year.
- US government agency estimates of the US oil endowment have grown over time such that US oil consistently could be produced for the next 35 years.

- US government agency estimates of the global oil endowment have grown over time such that global oil consistently could be produced for the next 45 years.
- Global oil reserves have more than doubled since 1980. From 1980 to 2008, the ratio of reserves to consumption has grown from 28 to 43. Global oil reserves have increased by 30 percent since 2000.
- OPEC's sudden jump (between 1988 and 1990) in its estimated reserves does not necessarily mean that it has exaggerated its available oil. From 1981 to 1996, reserve growth of giant oilfields in non-OPEC countries was three times that of OPEC reserve growth.
- Between 1986 and 2003, there was a significant lull in spending on oil exploration by major energy companies reporting to the US Department of Energy.
- The Middle East, Eastern Europe, and Africa contain three-quarters of world oil reserves and yet account for only one-seventh of exploratory drilling. Exploration has remained overly focused on North America.
- The success rate of US wells drilled for oil and gas increased from 20 percent in 1950 to over 50 percent in 2007.
- The US today uses less than half of the oil per unit of GDP than it did in 1975.
- China's use of oil to produce a unit of GDP has declined by two-thirds since 1980.
- Worldwide, the oil required to produce a unit of gross world product has decreased by almost 40 percent since 1980.
- Relative to the gross world product, the world spends less than half as much on oil today as it did in 1980, suggesting less dependence on oil by the global economy.
- Fifty percent more of US disposable income was spent on gasoline in 1980 than in 2008.
- Global supplies of heavy oil, oil sand, and oil shale greatly exceed the estimated global conventional oil endowment.
- At sustained oil prices of at least $80 (2007$) per barrel, the combination of remaining global conventional and unconventional oil is 4.5 times all the oil produced to date.
- Only 20 percent of the combined global endowment of oil and natural gas has been produced.
- Substitutes for oil are natural gas and coal, both of which can be converted into diesel, which is a preferred transportation fuel in much of the world. The estimated endowment of natural gas (in barrels of oil equivalent) that remains to be produced is greater than all remaining conventional oil. Coal supplies are plentiful and are outside of OPEC's control.

Notes and References

1. Deming, D. (2003). "Are We Running Out of Oil?" *Policy Backgrounder No. 159*, The National Center for Policy Analysis, Dallas, TX and Washington DC.
2. Hubbert, M. K. (1956). "Nuclear energy and the fossil fuels," presented at the Spring Meeting of the Southern District Division of Production, American Petroleum Institute, San Antonio, TX, March 1956: Shell Development Company Publication No. 95.
3. Deming, D. (2000). "Oil: Are We Running Out?" Second Wallace E. Pratt Memorial Conference, Petroleum Provinces of the 21st Century, January 12–15, 2000, San Diego, California.
4. Deming, D. (2003). "Are We Running Out of Oil?" *Policy Backgrounder No. 159*, The National Center for Policy Analysis, Dallas, TX and Washington DC.
5. Hall, C. A. S., and C. J. Cleveland (1981). "Petroleum Drilling and Production in the United States: Yield per Effort and Net Energy Analysis," *Science*, **211**, February 6, 1981: 576–9.
6. Deming, D. (2000). "Oil: Are We Running Out?" Second Wallace E. Pratt Memorial Conference, Petroleum Provinces of the 21st Century, January 12–15, 2000, San Diego, California.
7. Hubbert, M. K. (1982). "Techniques of prediction as applied to the production of oil and gas," in S. I. Gass (ed.), *Oil and Gas Supply Modeling*, National Bureau of Standards Special Publication **631**: 16–141.
8. See Deming, D. (2000). "Oil: Are We Running Out?" Second Wallace E. Pratt Memorial Conference, Petroleum Provinces of the 21st Century, January 12–15, 2000, San Diego, California.
9. Hubbert, M. K. (1982). "Techniques of prediction as applied to the production of oil and gas," in S. I. Gass (ed.), *Oil and Gas Supply Modeling*, National Bureau of Standards Special Publication 631: 139.
10. See Linden, H. J. (2004). "Rising expectations of ultimate oil, gas recovery to have critical impact on energy, environmental policy – Part 1," *Oil and Gas Journal*, **102**(3): 18.
11. Department of Energy (2006). *Undeveloped Domestic Oil Resources: The foundation for increasing oil production and a viable domestic oil industry*, prepared by Advanced Resources International, February 2006.
12. Lynch, M. (2003). "Petroleum resources pessimism debunked in Hubbert model and Hubbert modelers' assessment," *Oil and Gas Journal*, **101**(27), July 14, 2003.
13. Charpentier, R. R. (2005). "Estimating undiscovered resources and reserve growth: contrasting approaches," in A. G. Dore and B. A. Vining (eds), *Petroleum Geology: North-West Europe and Global Perspectives – Proceedings of the 6th Petroleum Geology Conference*, Petroleum Geology Conferences Ltd., Published by the Geological Society, London.
14. Duncan, R. C. (2001). "Energy resources—cornucopia or empty barrel?: Discussion," *AAPG Bulletin*, **85**(6), June 2001: 1090–2.

15. Deffeyes, K. S. (2005). *Beyond Oil: The View from Hubbert's Peak*. Farrar, Straus and Giroux.

16. Brandt, A. R. (2007). "Testing Hubbert," *Energy Policy*, **35**: 3074–88.

17. Ibid.

18. Deming, D. (2000). "Oil: Are We Running Out?" Second Wallace E. Pratt Memorial Conference, Petroleum Provinces of the 21st Century, January 12–15, 2000, San Diego, California.

19. Hubbert, M. K. (1956). "Nuclear energy and the fossil fuels," presented at the Spring Meeting of the Southern District Division of Production, American Petroleum Institute, March, 1956; graph and prediction reproduced in 1962 report *Energy Resources*, a report to the Committee on Natural Resources, National Academy of Sciences, National Research Council publication 1000-D, 141 pp.

20. Ibid.

21. Ibid.

22. Hubbert, M. K. (1971). "The Energy Resources of the Earth," *Scientific American*, in compilation *Energy and Power*: 31–40. Note: 1.35 trillion and 2.1 trillion barrel global oil endowment values appear. In 1975, Hubbert retreated from his global predictions of 1956 and 1971, stating: "I have not personally made world estimates for petroleum, …" (p. 17), "Methods and models for assessing energy resources," First IIASA Conference on Energy Resources, May 20–21, 1975 (ed. Michel Grenon, Pergamon Press, New York).

23. Maugeri, L. (2006). *The Age of Oil: The Mythology, History, and Future of the World's Most Controversial Resource*. Greenwood Publishing Group, Incorporated: 204.

24. From *Oil and Gas Journal* data, 2009; also appearing in EIA database.

25. Robelius, F. (2007). "Giant Oil Fields – The Highway to Oil," Doctoral dissertation, Uppsala University, Sweden: 75.

26. *Oil and Gas Journal*, "Worldwide look at reserves and production," Pennwell, December 24, 2007 (http://downloads.pennnet.com/pnet/surveys/ogj/071224ogj_24-25.pdf) and December 22, 2008 (http://downloads.pennnet.com/pnet/surveys/ogj/081222ogjwwlookatresandprod.pdf)

27. Hubbert, M. K. (1959). "Techniques of prediction with application to the petroleum industry" (preprint March 1959), Shell Development Company Publication 204, Houston, TX, 42 pp.

28. US Geological Survey, Mineral Commodity Summaries, January 2009: Reserve base is defined as "That part of an identified resource that meets specified minimum physical and chemical criteria related to current mining and production practices, including those for grade, quality, thickness, and depth. The reserve base is the in-place demonstrated (measured plus indicated) resource from which reserves are estimated."

29. US Geological Survey, Mineral Commodity Summaries, 2008: Reserve base value from USGS in 2008 for iron lead, and zinc.

30. Tierney, J. (1990). "Betting the Planet," *The New York Times*, December 2, 1990.
31. John Harte and John Holdren, physicists then both at the University of California, Berkeley, were also party to the bet: J. Giles, "Scientific wagers: Wanna bet?" *Nature* 420, November 28, 2002: 354–5, doi:10.1038/420354a. In March 2009, John Holdren was approved as Assistant to the President for Science and Technology (*The New York Times*, March 12, 2009).
32. US Census Bureau, www.census.gov/ipc/www/worldpop.html
33. Tierney, J. "Betting the Planet," *The New York Times*, December 2, 1990; "The $10,000 Question," *The New York Times*, August 23, 2005.
34. Brown, H. (1970). "Human materials production as a process in the biosphere," *Scientific American*, **223**(3), September 1970: 195–208. Brown had noted that plastics were rapidly serving as substitutes for metals but did not imagine their role in the history of copper: "The use of synthetic plastics is now increasing with impressive speed. Total world production of these materials now exceeds in both volume and weight the production of copper and aluminum combined." (p. 205).
35. http://minerals.usgs.gov/minerals/pubs/commodity/copper/mcs-2009-coppe. pdf. Interestingly, the USGS 2008 reserve estimate is 550 million metric tons, which again points to about 35 years of supply at the 2008 global production rate of 15.7 million metric tons per year.
36. US Geological Survey, Mineral Commodity Summaries, January 2008, Copper.
37. Estimates based on projection of data presented in "Sustainable Development of the Copper Sector: A Global View," Patrick Hurens, Secretary General, © 2004 International Copper Study Group 1, China International Copper Forum, October 27–29, 2004, Haikou, China.
38. Brooks, W. E. (2006). *2005 Minerals Year Book*, Mercury. US Geological Survey; USGS (2007). Mineral Commodity Summary, page 104, Mercury; USGS (2000). Fact Sheet 146-00, Mercury in the Environment.
39. A generalized resource production figure of this sort was presented in McCabe (1998). McCabe, P. J. (1998). "Energy resources – cornucopia or empty barrel?" *AAPG Bulletin*, **82**(11): 2110–34.
40. The value is 37.6 billion barrels based on a population of 9.22 billion times 4.08 barrels per person per year: UN (2004). *World Population to 2300*, Department of Economic and Social Affairs, ST/ESA/SER.A/236, 240 pp., New York (see p. 12).
41. Hall, C. A. S., and C. J. Cleveland (1981). "Petroleum Drilling and Production in the United States: Yield per Effort and Net Energy Analysis," *Science*, **211**(6), February 1981: 576–9.
42. Imports between 1955 and 1970 averaged 12.6 percent of the sum of production and imports; 26.5 percent in 1973; reached 33.4 and 39.9 percent in 1975 and 1976: and between 1971 and 1986, the average was 34.5 percent.
43. Rifai, T. (1974). *Pricing of Crude Oil: Economic and strategic guidelines for an international energy policy*. New York: Praeger, 138. Original source: *World Tanker Fleet Review*. London: John I. Jacobs and Co., 12/31/72.

44. Held, C. (2000). *Riches Beneath the Earth, from Middle East Patterns. People, Places and Politics*. Westview Press, Perseus Books Group; Rifai (1974), ibid.; Yergin, D., (1991). *The Prize: The Epic Quest for Oil, Money, and Power*. New York: Free Press, Simon and Schuster.
45. Rifai, T. (1974). *Pricing of Crude Oil: Economic and strategic guidelines for an international energy policy*. New York: Praeger.
46. Porter, E. D. (1995). "Are We Running Out of Oil?" American Petroleum Institute, Policy Analysis and Strategic Planning Department, Discussion Paper #081.
47. Porter (1995), ibid., citing Stauffer, T. [1994a]. "Trends in Oil Production, Costs in the Middle East, and Elsewhere," *Oil and Gas Journal*, March 21: 105–7.
48. This approach to adjusting for inflation is suggested by Julian Simon in *The Ultimate Resource: People, Materials, and Environment* (published by Princeton University Press, 1981).
49. Oil price data, *BP Statistical Review of World Energy, 2008*; inflation adjustment, R. Sahr (2009), http://oregonstate.edu/cla/polisci/faculty-research/sahr/sahr.htm
50. Williamson, S. H. (2008). "Six Ways to Compute the Relative Value of a U.S. Dollar Amount, 1774 to present," www.measuringworth.com/uscompare/#
51. Sahr, R. (2009). Inflation Conversion Factors for Dollars 1774 to Estimated 2018, http://oregonstate.edu/cla/polisci/faculty-research/sahr/sahr.htm
52. Fisher, A. C. (1981). *Resource and Environmental Economics*. Cambridge: Cambridge University Press, 284 pp.
53. As discussed in Livernois and Martin (2003), one problem with using the idea of scarcity rent is that it may not indicate much about scarcity. Economists disagree on whether scarcity rent is expected to increase or decrease as a non-renewable resource is depleted. The gap between price and extraction cost can be narrowed when extraction costs rise. Indeed, higher and higher extraction costs would be expected as reserves become more difficult to produce (e.g. expensive deep offshore wells, or small, isolated reservoirs). Ultimately, the cost of extraction can increase to the point that the price does not justify the effort to further mine the resource; at that point a cheaper alternative, or backstop, would presumably become viable. Livernois, J., and P. Martin (2003). "Price, scarcity rent, and a modified r per cent rule for non-renewable resources," *Canadian Journal of Economics*, **34**(3): 827–45.
54. Hamilton, J. D. (2009). "Understanding crude oil prices," *Energy Journal*, **30**(2), based on NBER Working Paper No. 14492, issued in November 2008.
55. Fisher, A. C. (1981). *Resource and Environmental Economics*. Cambridge: Cambridge University Press, 284 pp.
56. Energy Information Administration, *International Energy Outlook 2004 – World Oil Markets*, www.eia.doe.gov/oiaf/archive/ieo04/oil.html; EIA Financial Reporting System (2008) – note that average US total finding and production cost was $17 to $25 between 2005 and 2006.

57. McCabe, P. J. (1998). "Energy resources—cornucopia or empty barrel?" *AAPG Bulletin*, **82**(11): 2110–34.

58. Used here are the Department of Energy 2006 estimate of US technologically available oil as the latest indication of the oil endowment and cumulative production data from EIA.

59. McCabe (1998). "Energy resources—cornucopia or empty barrel?" *AAPG Bulletin*, **82**(11): 2110–34.

60. Department of Energy (2006). "Undeveloped domestic oil resources: The foundation for increasing oil production and a viable domestic oil industry," prepared by Advanced Resources International.

61. McCabe (1998). "Energy resources—cornucopia or empty barrel?" *AAPG Bulletin*, **82**(11): 2110–34.

62. Weeks, L. G. (1948). "Highlights on 1947 developments in foreign petroleum fields," *AAPG Bulletin*, **32**: 1093–1160.

63. Hubbert, M. K. (1956). "Nuclear energy and the fossil fuels," Shell Development Company Publication 95.

64. Salvador, A. (2005). "Energy: A historical perspective and 21st century forecast," American Association of Petroleum Geologists, Studies in Geology No. 54.

65. McCabe, P. J. (1998). "Energy resources—cornucopia or empty barrel?" *AAPG Bulletin*, **82**(11): 2110–34.

66. Times are calculated by following a similar approach to that of McCabe (1998), ibid., but using the consumption rate at the time the reserve estimates were made; McCabe considered the actual consumption rate over the hypothetical "depletion period." Values are based on EIA data, but almost identical results are obtained using values published by BP. It should also be noted that if production and not consumption data are used, the average reserve-to-production ratio is about 10 percent greater.

67. Salvador, A. (2005). "Energy: A historical perspective and 21st century forecast," American Association of Petroleum Geologists, *Studies in Geology*, No. **54**: 46, Table 8.

68. Lynch, M. (2003). "Petroleum resources pessimism debunked in Hubbert model and Hubbert modelers' assessment," *Oil and Gas Journal*, July 14, 2003; Maugeri, L. (2004). "Oil: Never Cry Wolf—Why the Petroleum Age Is Far from Over," *Science*, **304**(21), May 2004: 1114–15.

69. Campbell, C. J. and J. H. Laherrère (1998). "The end of cheap oil," *Scientific American*, March 1998: 78–83.

70. Campbell, C. J. (2002). "Forecasting Global Oil Supply 2000–2050," Hubbert Center Newsletter # 2002/3, M. King Hubbert Center for Petroleum Supply Studies, on page 5: "The base-case scenario points to 2010, but it could come sooner if economic recovery should drive up the demand for oil."

71. Campbell, C. J. and J. H. Laherrère (1998). "The end of cheap oil," *Scientific American*, March 1998: 78–83.

72. Charpentier (2005), p. 6, citing Klett et al. (2000) and Klett and Schmoker (2003). Charpentier, R. R. (2005). "Estimating undiscovered resources and

reserve growth: contrasting approaches," in A. G. Dore and B. A. Vining (eds), *Petroleum Geology: North-West Europe and Global Perspectives – Proceedings of the 6th Petroleum Geology Conference*, 3–9. Petroleum Geology Conferences Ltd., published by the Geological Society, London; Klett, T. R., R. R. Charpentier, J. W. Schmoker, and E. D. Attanasi (2000). "Predicting changes in world oil and gas field sizes." Abstract, presented at the American Association of Petroleum Geologists Annual Meeting, New Orleans, LA, April 16–19, A79; Klett, T. R., and J. W. Schmoker (2003). "Reserve growth of the world's giant oil fields," in M. T. Halbouty (ed.), *Giant oil and gas fields of the decade 1990–1999*. American Association of Petroleum Geologists Memoir, 78: 107–22.

73. Discussion based on Demirmen (2004) and Securities and Exchange Commission case filing. Demirmen, F. (2004). "Shell's reserve revision: A critical look," *Oil and Gas Journal*, April 5, 2004: 43–6.

74. United States of America Before the Securities and Exchange Commission Securities Exchange Act of 1934, Release No. 50233/August 24, 2004, Accounting and Auditing Enforcement Release No. 2085/August 24, 2004, Administrative Proceeding File No. 3-11595, In the Matter of Royal Dutch Petroleum Company and The "Shell" Transport and Trading Co., plc. Respondents.

75. BBC News, "Shell settles oil reserve claims," April 11, 2007; "Shell makes record profits," February 3, 2006; "Shell gets surprise profit boost," May 3, 2007.

76. http://money.cnn.com/magazines/fortune/global500/2006/snapshots/1154.html

77. "Shell reports 33% rise in profit," *International Herald Tribune*, July 31, 2008, www.iht.com/articles/2008/07/31/business/31shellNEW.php

78. Energy Information Administration, 2009.

79. Curtiss, D. (2009). "Reserves Disclosure Rules Revised," *AAPG Explorer*, February 2009, www.aapg.org/explorer/washingtonwatch/2009/02feb.cfm

80. EIA data based on *Oil and Gas Journal*, January 2008 and 2009.

81. Bahorich, M. (2006). "End of oil? No it's a new day dawning," *Oil and Gas Journal*, **104**(31), August 21, 2006: 30–4.

82. McCabe, P. J. (1998). "Energy resources—Cornucopia or empty barrel?" *AAPG Bulletin*, **82**(11), 2110–134.

83. Ibid.

84. US Minerals Management Service (2006). "Assessment of Undiscovered Technically Recoverable Oil and Gas Resources of the Nation's Outer Continental Shelf, 2006."

85. Conversion: 1 million cubic feet of natural gas is 172.3 barrels of oil equivalent.

86. Pinsker, L. M. (2003). "Raining hydrocarbons in the Gulf," *Geotimes*: www.geotimes.org/june03/NN_gulf.html

87. The DOE total is 1.332 trillion barrels, of which 80 billion barrels are oil sands (not counted here).

88. DOE (2005). "Performance Profiles of Major Producers 2004"; DOE (2008). "Performance Profiles of Major Producers 2007."

89. The period from 2002 to 2005 averaged only about 70 percent, falling from over 115 percent during the period from 1999 to 2002.

90. DOE (2008). "Performance Profiles of Major Energy Producers 2007."

91. Klett, T. R. (1998). "Graphical Comparison of Reserve-Growth Models for Conventional Oil and Gas Accumulations," Chapter F of *Geologic, Engineering, and Assessment Studies of Reserve Growth*, edited by T. S. Dyman, J. W. Schmoker, and M. Verma, U.S. Geological Survey Bulletin 2172–F, US Department of the Interior, US Geological Survey.

92. Note that known versus grown oil volumes were computed and volume adjustments were made to account for discovery reporting dates before 1996, which is the beginning date of the USGS's 30-year assessment time-frame.

93. Schmoker, J. W. and T. R. Klett (2000). "Estimating potential reserve growth of known (discovered) fields—A component of the USGS World Petroleum Assessment 2000," in US Geological Survey World Energy Assessment Team, *US Geological Survey World Petroleum Assessment 2000*: US Geological Survey Digital Data Series DDS-60, 4 CD-ROMs, 20 p.

94. Klett, T. R., D. L. Gautier, and T. A. Ahlbrandt (2005). "An evaluation of the US Geological Survey World Petroleum Assessment 2000," *AAPG bulletin*, **89**(8), 1033–42.

95. www.eia.doe.gov/emeu/perfpro/pp05overview.ppt, PowerPoint presentation values and 2006 value from 2007 EIA FRS report.

96. Horn, M. K. (2009). www.sourcetoreservoir.com. Note that the estimated volumes of ultimately recoverable condensate from natural gas are based on reservoir calculations, including hydrocarbon dew point temperatures and chemistry of the raw gas (M. Horn, personal communication, 2009).

97. "New Brazil discovery may be giant," *Oil and Gas Journal*, April 14, 2008; Clendenning, A. (2008). "Brazil oil field could be huge find," Associated Press, April 14, 2008.

98. Horn, M. K. (2009). www.sourcetoreservoir.com

99. Sandrea, R. (2006). "Early new field production estimation could assist in quantifying supply trends," *Oil and Gas Journal*, **104**(20), May 22, 2006.

100. Horn, M. K. (2006). "Giant fields 1868–2003," data on CD-ROM, in M. Halbouty (ed.), "Giant oil and gas fields of the decade 1990–1999," AAPG Memoir 78, 2003, 340 pp., modified November 2006 to reflect giant oil-field discoveries 2000 to 2006.

101. "Major discovery increases oil reserve," *Xinhua/China Daily*, May 3, 2007.

102. Izundu, U. (2008). "Special Report: Ghana due first oil output in 2010 with Jubilee start-up," *Oil and Gas Journal*, **106**(20), May 26, 2008.

103. Sandrea, I., and R. Sandrea (2007). "Exploration trends show continued promise in world's offshore basins," *Oil and Gas Journal*, **105**(9), March 5, 2007: 34–40.

104. Sandrea, R. (2006). "Early new field production estimation could assist in quantifying supply trends," *Oil and Gas Journal*, **104**(20), May 22, 2006.

105. Bahorich, M. (2006). "End of oil? No it's a new day dawning." *Oil and Gas Journal*, **104**(31), August 21, 2006: 30–4.

106. Sandrea, I., and M. Barkindo (2007). "West Africa – 1: Undiscovered oil potential still large off West Africa," *Oil and Gas Journal*, **105**(2), January 8, 2007.

107. Mufson, S. (2006). "U.S. Oil Reserves Get a Big Boost, Chevron-Led Team Discovers Billions of Barrels in Gulf of Mexico's Deep Water," *Washington Post*, Wednesday, September 6, 2006.

108. "Chevron makes new oil discovery in Gulf of Mexico," Reuters (UK), February 5, 2009.

109. Charpentier, R. R. (2005). "Estimating undiscovered resources and reserve growth: contrasting approaches," in A. G. Dore and B. A. Vining (eds), *Petroleum Geology: North-West Europe and Global Perspectives—Proceedings of the 6th Petroleum Geology Conference, 3–9*. Petroleum Geology Conferences Ltd, published by the Geological Society, London.

110. Charpentier (2005), ibid.: 4–6.

111. Fletcher, S. (2006). "Facts: Oil market suffering from investment lull of 1990s," *Oil and Gas Journal*, October 16, 2006: 18–20.

112. Horn (2009). www.sourcetoreservoir.com

113. Baker, D. R. (2007). "Big Oil cautious about clean-energy spending, Critics want more from firms earning billions," staff writer, *San Francisco Chronicle*, February 9, 2007; Mufson, S. (2007). "Higher Oil Prices Help Exxon Again Set Record Profit," *Washington Post*, February 2, 2007; "Chevron Announces $22.8 Billion Capital and Exploratory Budget for 2009," www.chevron.com/News/Press/release/?id=2009-01-29; "Chevron Announces $22.9 Billion Capital and Exploratory Budget for 2008." www.chevron.com/news/press/release/?id=2007-12-06.

114. "Exxon Mobil: A Great Big Buy," June 3, 2008, www.businessweek.com/investor/content/jun2008/pi2008063_252790.htm; Hargreaves, S. (2009). "Exxon 2008 profit: A record $45 billion," CNNmoney.com, January 30, 2009; Gold, R. (2009). "Exxon Mobil to pump up cash reserves as profit falls," *Wall Street Journal*, May 1, 2009. Note that Chevron ended its stock buy-back program in 2009: Carroll, J. (2009). "Exxon Targets Nine New Projects in 2009 Spending Plan," Bloomberg.com, March 5, 2009.

115. "World giant oil discoveries seem not to be at an end," news item, *Oil and Gas Journal*, November 6, 2006: 33.

116. Barkindo, M. and I. Sandrea (2007). "Undiscovered oil potential still large off West Africa," *Oil and Gas Journal*, **105**(2), January 8, 2007: 30–4.

117. Africa: $0.91 billion in 2000 compared with $2.1 billion in 2005 (nominal dollars).

118. Middle East: $0.056 billion in 2000 compared with $0.31 billion in 2005 (nominal dollars).

119. Berman, A. (2005). "Ideas are like stars: The current oil boom, 2001–2005," *Houston Geological Society Bulletin*, June 2005. Also, see Berman, A. (2004). "Oil and gas reserves reduction – A geologist's perspective," *Houston Geological Society Bulletin*, May 2004.

120. www.eia.doe.gov/emeu/cabs/Russia/Background.html and IMF study cited there.

121. Ulmishek, G. F. (2003). "Petroleum Geology and Resources of the West Siberian Basin, Russia," USGS Bulletin 2201-G.

122. Ulmishek (2003), ibid.

123. US Geological Society (2008). "Circum-Arctic Resource Appraisal: Estimates of Undiscovered Oil and Gas North of the Arctic Circle," USGS Fact Sheet 2008–3049; "90 Billion Barrels of Oil and 1,670 Trillion Cubic Feet of Natural Gas Assessed in the Arctic," US Geological Society News Release, July 23, 2008.

124. Consumption data from EIA; population data from Economic Research Service, USDA.

125. Income approximated as per capita GDP; GDP based on GDP Purchasing Power Parity (PPP) is $6,200 per capita in China versus $41,500 per capita in the US.

126. EIA 2008 and *BP Statistical Review of World Energy*, June 2008. The Russian Federation was formed in December 1991, and 1992 is the first full year for comparison. However, in 1985 the Russian Federation, before dissolution of the USSR, consumed about half of the oil it consumed in 2007.

127. Note: 1980 Russia GDP was estimated as 41 percent of FSU GDP in 1980; 41 percent is the five-year average value of Russian versus FSU GDP from 1985–1989.

128. World Development Indicators database, World Bank, 1 July 2008; http://siteresources.worldbank.org/DATASTATISTICS/Resources/GDP.pdf and www.cia.gov/library/publications/the-world-factbook/geos/xx.html#Econ

129. *International Monetary Fund World Economic Outlook*, 2006, Chapter II, Oil Prices and Global Imbalances, p. 73.

130. www.globalfinancialdata.com/articles/Oil_Is_History_Repeating_Itself.doc (2006).

131. Hooker, M. A. (2002). "Are oil shocks inflationary? Asymmetric and nonlinear specifications versus change in regime," *Journal of Money, Credit, and Banking*, **34**(2), May 2002: 540–61.

132. Walton, D. (2006). "Has oil lost the *capacity* to shock?" *Oxonomics*, **1**(2006): 9–12.

133. Kirkemo, H., W. L. Newman, and R. P. Ashley (1997). "Gold," US Geological Survey Report.

134. World Gold Council, gold.org/discover

135. www.usgs.gov/faq/list_faq_by_category/get_answer.asp?id=89; volume calculation here based on density of 19.3 grams per cubic centimeter.

136. http://minerals.usgs.gov/minerals/pubs/mcs/2009/mcs2009.pdf. From 1900 through 2007, 131,000 metric tons of gold were mined globally.

137. The "reserve base" of 100,000 metric tons or 3.2 billion troy ounces is defined as encompassing "those parts of the resources that have a reasonable potential for becoming economically available within planning horizons beyond those that assume proven technology and current economics. The reserve base includes those resources that are currently economic (reserves), marginally economic (marginal reserves), and some of those that are currently subeconomic (subeconomic resources)." USGS Mineral Commodities Summary, 2009, http://minerals.usgs.gov/minerals/pubs/mcs/2009/mcs2009.pdf and http://minerals.usgs.gov/minerals/pubs/commodity/gold/mcs-2008-gold.pdf

138. Butterman, W. C. and E. B. Amey III (2005). "Mineral Commodity Profiles— Gold," USGS Open-File Report 02-303.

139. www.eoearth.org/article/Gold

140. Rex Tillerson, in a speech to Boston CEO Club, November 30, 2006, www.edf.org/article.cfm?contentid=5691

141. US DOE (2006). "Undeveloped domestic oil resources: The foundation for increased oil production and a viable domestic oil industry," prepared for the DOE by Advanced Resources International, February 2006.

142. Ibid.

143. International Energy Agency (2005). *Resources and Reserves, Oil and Gas Technologies for the Energy Markets of the Future*, IEA Publications, printed in France by JOUVE (61 2005 25 1 P1), ISBN 92-64-109-471 – 2005.

144. Sweetnam, G. (2008). "Long-term Global Oil Scenarios: Looking Beyond 2030," Energy Information Administration, April 7, 2008 Energy Conference, Washington DC. EIA estimates based on I.H.S. Energy, World Energy Council, USGS, Nehring Associates, and EIA analysis.

145. US DOE (February 2006). "Project Facts, Game Changer Improvements Could Dramatically Increase Domestic Oil Recovery Efficiency," http://fossil.energy.gov/programs/oilgas/publications/eor_co2/D_-_PROJECT_FACT_GAME_CHANGER.pdf

146. In DOE (2006), based on 2004$ conversion of information in Birol, F., and W. Davie (2001). "Oil Supply Costs and Enhanced Productivity," *Energy Prices and Taxes* (4th Quarter 2001), pp. xvii–xxii, http://data.iea.org/ieastore/assets/products/eptnotes/feature/4Q2001B.pdf

147. Maugeri, L. (2004). "Oil: Never Cry Wolf—Why the Petroleum Age Is Far from Over," *Science*, **304**, May 21, 2004: 1114–15.

148. Meyer, R. F. and E. D. Attanasi (2003). "Heavy Oil and Natural Bitumen – Strategic Petroleum Resources," USGS fact sheet 70-03 (excludes FSU Siberian extra-heavy oil).

149. Salvador, A. (2005). "Energy: A historical perspective and 21st century forecast," *AAPG Studies in Geology*, No. **54**.

150. Meyer, R. F. and E. D. Attanasi (2003). "Heavy Oil and Natural Bitumen – Strategic Petroleum Resources," USGS fact sheet 70-03 (excludes FSU Siberian extra-heavy oil).

151. Caruso, G. (2005). "When will world oil production peak?" Guy Caruso, Administrator, Energy Information Administration, US Department of Energy, 10th Annual Asia Oil and Gas Conference, Kuala Lumpur, Malaysia. June 13, 2005.

152. EIA Venezuela country brief – Oil, September 2005.

153. Alberta Energy and Utilities Board (2004). "Alberta's Reserves 2003 and Supply/Demand Outlook 2004–2013," Statistical Series (ST) 2004-98, Calgary, Alberta, June 2004.

154. Ahlbrandt, T. S., and P. J. McCabe (2002). "Global Petroleum Resources: A View to the Future," *Geotimes*, November 2004.

155. Caruso, G. (2005). "When will world oil production peak?" Guy Caruso, Administrator, Energy Information Administration, US Department of Energy, 10th Annual Asia Oil and Gas Conference, Kuala Lumpur, Malaysia, June 13, 2005.

156. EIA, www.eia.doe.gov/oiaf/aeo/issues.html

157. Canadian Association of Petroleum Producers, www.capp.ca/default.asp? V_DOC_ID=603

158. Park, G. (2008). "Alberta oil sands frenzy fizzling," *Petroleum News*, **13**(51), December 21, 2008: 8; "Study sees pause in oil sands output growth," *Oil and Gas Journal*, February 16, 2009: 5.

159. Rand Report (2005). Bartis, J. T., T. LaTourrette, L. Dixon, D. J. Peterson, and G. Cecchine (2005). *Oil Shale Development in the United States: Prospects and Policy Issues*, Rand Corporation, Monograph series.

160. Ibid.

161. Ibid.

162. Biglarbigi, K., H. Mohan, M. Carolus, and J. Killen (2009). "Analytic approach estimates oil shale development economics," *Oil and Gas Journal*, February 2, 2009: 48–53.

163. "Israel presses for oil from shale," *Business Week*, July 5, 2006.

164. Dyni, J. R. (2003). "Geology and resources of some world oil-shale deposits," *Oil Shale*, **20**(3): 193–252.

165. EIA (2006). *World Energy Outlook*.

166. Stranges, A. N. (2003). "Germany's Synthetic Fuel Industry 1927–45," Prepared for Presentation at the AIChE 2003 Spring National Meeting, New Orleans, LA, March 30–April 3, 2003, Historical Development of the Fischer-Tropsch Synthesis Process – I.

167. Rahmin, I. I. (2005). "Stranded gas, diesel needs push GTL work," *Oil and Gas Journal*, March 14, 2005: 18–28.

168. Jager, B. (2003). "Development of Fischer-Tropsch Reactors," Prepared for Presentation at the AIChE 2003 Spring National Meeting, New Orleans, LA, March 30–April 3, 2003, Historical Development of the Fischer-Tropsch Synthesis Process – II.

169. www.sasolchevron.com/fischer-tropsch_conversion.htm

170. Bullis, K. (2006). "Clean Diesel from Coal," *MIT Technology Review*, April 19, 2006; Goldman, A. S., A. H. Roy, Z. Huang, R. Ahuja, W. Schinski, and

M. Brookhart (2006). "Catalytic Alkane Metathesis by Tandem Alkane Dehy-drogenation – Olefin Metathesis," *Science*, **312**(14), April 2006.

171. Polk Automotive Intelligence, R. L. Polk & Company, 2006.
172. "Rudolf's Revenge," *The Economist*, February 9, 2007. (Note: The standard was slated to go into effect in 2009.)
173. US DOE (2008). www1.eere.energy.gov/vehiclesandfuels/facts/2007_fcvt_fotw481.html, and IFP Panorama Technical Report (2005), *Road Transport Fuels in Europe: the Explosion of Demand for Diesel Fuel.*
174. Rahmin, I. I. (2005). "Stranded gas, diesel needs push GTL work," *Oil and Gas Journal*, March 14, 2005: 18–28.
175. www.energy.ca.gov/afvs/synthetic_diesel.html
176. Dittrick, P. (2007). "Military testing Fischer-Tropsch Fuels," *Oil and Gas Journal*, February 26, 2007: 24–5.
177. Sasol, Inc., Press Release, "Qatar Petroleum and Sasol sign Joint Venture Agreement for the Pioneer GTL Plant in Qatar," July 10, 2001: www.sasolchevron.com/pr_04.htm
178. Robinson, K. K. and D. F. Tatterson (2007). "Fischer-Tropsch oil-from-coal promising as transportation fuel," *Oil and Gas Journal*, February 26, 2007: 20–31.
179. Sasol, Inc., Press Release, "Sasol and Chinese sign landmark coal-to-liquids agreement," June 22, 2006.
180. Miller, C. L. (March 2007). "Coal Conversion – Pathway to Alternate Fuels," 2007 EIA Energy Outlook Modeling and Data Conference, Washington DC.
181. Rahmim, I. I. (2008). "Special Report: GTL, CTL finding roles in global energy supply," *Oil and Gas Journal*, **106**(12), March 24, 2008.
182. "Coal-to-oil project approved," China Daily, January 11, 2008.
183. EIA (2007). *World Energy Outlook.*
184. "China May Halt Coal-to-Oil Projects," *The New York Times*, June 10, 2007.
185. Rahmin, I. I. (2005). "Stranded gas, diesel needs push GTL work," *Oil and Gas Journal*, **103**(10), March 14, 2005.
186. www.eia.doe.gov/pub/international/iea2005/table41.xls
187. The energy equivalent of natural gas 5.7 quadrillion cubic feet of gas is approximately 1 trillion barrels of oil.
188. Klett, T. R., D. L. Gautier, and T. S. Ahlbrandt (2005). "An evaluation of the U.S. Geological Survey World Petroleum Assessment 2000," *AAPG Bulletin*, **89**(8): 1022–42.
189. Ahlbrandt, T. S. and P. J. McCabe (2002). "Global Petroleum Resources: A View to the Future," *Geotimes*, November 2002.
190. 1 metric ton (or tonne) of coal is equal to 2204.6 pounds, which is equivalent to 3.8 barrels of oil.
191. 2008 *BP Statistical Review of World Energy* and 2006 World Energy Council Survey of Energy Resources, www.worldenergy.org
192. Lower estimate from 2008 *BP Statistical Review* and the higher from EIA 2007 *World Energy Outlook.*

5

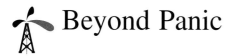 Beyond Panic

High gasoline prices in 2008 served as a reminder of our vulnerability to limited domestic oil production and dependence on oil imports, especially from OPEC. The future of dependence on imports and, more generally, the oil-energy path requires public policy decisions about energy security, economic stability, and risks to the environment. These issues are discussed in this chapter. The US and other oil-importing nations must resolve the problems of energy dependence and environmental degradation due to oil consumption. Should global oil supply meet demand for the foreseeable future, we must ask: at what cost to our safety, economy, and environment?

The Non-Renewable Resource Model

Perhaps lessons from historical production of other globally traded non-renewable Earth resources are applicable to oil. More importantly, if there are such lessons, can they help inform us about how to move to a sustainable transportation-fuel future? For various non-oil commodities, production has often followed a bell-shaped curve. But decline to nominal production rates in the US has not meant that the US has effectively run out of natural commodities such as iron ore, zinc, and lead. In fact, based on historical production of non-oil commodities, peak production and decline have occurred even though plenty of the resource remains.

    Under most circumstances, a peak in production corresponds to a peak in consumption. The expectation that a resource will be depleted is often sufficient to drive a pattern of behavior that results in a peak in consumption,

Oil Panic and the Global Crisis: Predictions and Myths. 1st edition. By Steven M. Gorelick.
Published 2010 by Blackwell Publishing, ISBN 978-1-4051-9548-5 (hb)

which prompts a peak in production. Uncertainty in the marketplace can cause prices to rise because of a perception of scarcity. After an initial short-term reaction, such as hoarding of a commodity, which drives up prices further, the first responses to a price rise are greater efficiency in the use of the resource and softening of demand. If the price remains high for a sufficiently long period, innovation is spurred, leading to increases in availability of substitutes for the resource. Ultimately, demand for the original commodity drops in favor of a physical substitute, or there is a shift in consumer behavior to satisfy the end-use need in the absence of the original "essential" commodity. Will this be the case with oil?

We can start by considering that the behavioral pattern of consumer response, which has driven production of other Earth resources, is applicable to oil. To explore this idea, three questions must be answered:

- Where is an efficiency gain possible?
- Will increases in efficiency indeed reduce demand?
- What might ultimately substitute for oil?

Where Is an Efficiency Gain Possible?

Forty percent of US energy comes from oil and 22 percent from natural gas, with the remainder produced primarily from coal and hydropower. In the US, over two-thirds of oil is used for transportation fueled by motor gasoline, jet fuel, and diesel fuel (Figure 5.1). For over 20 years, the proportion has been similar, with more than 60 percent of crude oil used for transportation fuel.

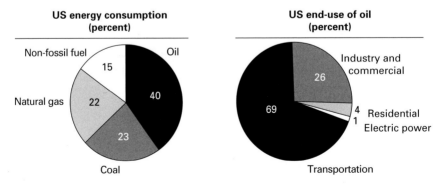

Figure 5.1   Oil provides the greatest share of US primary energy consumption, and oil is the major source of transportation energy. (Data: EIA)

Two-thirds of transportation fuel is for cars and light (four-wheel) trucks,[1] and about seven-eighths of transportation fuel is for cars, trucks, buses, and motorcycles, with the remaining one-eighth used to power airplanes, boats, and trains.

The greatest potential efficiency gains in our use of oil lie in transportation. During 2007, oil-propelled, four-wheel motor vehicles in the US traveled an astonishing 3 trillion miles.[2] To put this distance in perspective, it equals 6.3 million round trips from Earth to the moon, or 16,100 round trips from Earth to the sun. In 2008, US drivers used 138 billion gallons of gasoline at a cost of about $400 billion. With such a high rate of fuel consumption, vehicle fuel economy is a key target of opportunity.

Well into the 1970s, light-duty vehicles (cars and light trucks) attained less than 15 miles per gallon (mpg) on average. In today's terms, virtually every passenger car was a gas-guzzler, averaging between 13.5 and 14.5 miles per gallon between 1960 and 1975. Then something remarkable occurred. As a consequence of the 1973 oil crisis following the OPEC oil embargo, which limited supply and caused oil prices to rise, Congress passed into law the **Corporate Average Fuel Economy (CAFE) standards** in 1975 as part of the Energy Policy Conservation Act. The objective of CAFE was to double the average fuel efficiency of new cars in 1974 to 27.5 mpg (8.6 liters/100 kilometers) within 10 years from enactment of the CAFE standards through fuel economy benchmarks that phased in rapidly.

The result of implementing the new standards was that average gasoline mileage for the blend of new and old cars on the road increased by 50 percent from under 14 mpg in 1972 to over 21 mpg in 1991; the fuel economy of light trucks improved from under 10 mpg in 1972 to 17 mpg in 1991 (Figure 5.2).

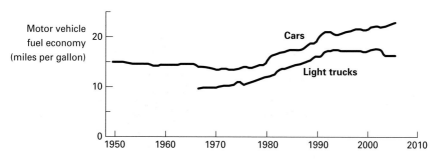

Figure 5.2   Fuel economy of passenger cars and light trucks (defined as SUVs, vans, and pickups under 8,500 pounds) on the road in the US from 1949 to 2005, showing the beneficial impact of the CAFE standards in producing about a two-thirds increase in miles per gallon since 1975. (Data: EIA)

Gasoline consumption per passenger car fell from 754 gallons in 1972 to 554 gallons in 2006 (from 922 to 690 gallons for light trucks, vans, and SUVs). This decline occurred even though Americans drove more. The annual mileage driven per vehicle in the US has increased by over 20 percent since 1972 (Figure 5.3 for passenger cars), from about 10,000 to over 12,000 miles per year.

The primary increase in car fuel efficiency occurred over a period of just about five years in the early 1980s, when US consumers rapidly bought new fuel-efficient cars. From 1970 to 1980, US sales averaged about 8.4 million new cars per year. New car sales increased to almost 11 million in 1985. The surge in new car purchases from 1980 to 1985 was equivalent to replacing 40 percent of the cars registered in the US. Older, fuel-inefficient cars were rapidly replaced by new cars that met the higher miles-per-gallon CAFE standards.

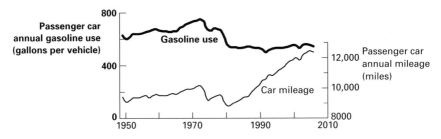

Figure 5.3   Fuel consumption of on-road passenger cars (left axis) has declined since the OPEC oil embargo of 1973 and the CAFE standards went into effect even though car annual mileage (right axis) increased by over 20 percent since that time. (Data: EIA)

How were these fuel efficiency standards met so rapidly? The answer was reducing vehicle weight. Weight is the key determinant of energy needed to propel a car.[3] In 1975, the average light-duty vehicle weighed over 4,000 pounds, but by 1981 that value had declined to 3,100 pounds.[4] The weight reduction was achieved by making smaller cars, substituting lighter aluminum, magnesium, and plastics for body parts, and using aluminum in the engines and axles. The average amount of steel in vehicles dropped by 700 pounds in just 10 years, from 3,160 pounds in model year 1970 to 2,460 pounds in 1980. By 2004, the average new car's steel content was just 2,150 pounds, a reduction of 1,000 pounds from 1970.[5]

Implementation of the CAFE standards showed that efficiency gains in the transportation sector significantly reduced the rate of consumption of gasoline. Improved automotive fuel efficiency resulted in an average increase from about 14 to 23 miles per gallon, which saved enormous amounts of oil and gasoline.

With no change to the CAFE standards since 1975, new car fuel economy plateaued in 1988 and has become slightly worse (5 percent) since then. In December 2007, the **Energy Independence and Security Act of 2007** became law. It raised the CAFE fuel efficiency standard for cars and light trucks to *35 miles per gallon by 2020*. This time there was no loophole for light trucks (discussed in the next section) as in the original CAFE standards. Moving from the existing standard of 27.5 mpg to 35 mpg over a dozen years seems rather modest, but nonetheless the legislative effort was vigorously opposed by the automotive industry, which lobbied for a 30 mpg standard.[6] Many cars on the road today have better fuel economy than the 35 mpg newly adopted standard, so there are apparently few technological impediments to meeting it, especially considering the 2020 time horizon for complete implementation.[7]

In comparison to the new US standard, Japan requires an average fuel economy of 44 mpg (as of 2008), China 34 mpg (as of 2005), and Australia and the European Union, in voluntary agreements with industry, expect 34 mpg (by 2010) and 44 mpg (as of 2008), respectively.[8] Of course, cars in European countries often are smaller and use more efficient diesel engines than in the US. In addition, there is a much higher proportion of European cars with manual transmission than in the US, although this does not produce the mileage advantage seen historically. Today manual transmission cars account for 80 percent of those produced in Europe and 70 percent produced in China. In North America, only 10 percent of new cars are produced with manual shifts.[9]

Will Increases in Efficiency Indeed Reduce Demand?

In 1865, British economist William Stanley Jevons claimed that increasing the efficiency of the use of a natural resource increases consumption. That is, greater efficiency does not produce the desired effect of reducing consumption. He argued that the more efficient use of coal, for example through improvements in the steam engine, decreased the cost of manufactured goods, which in turn raised demand for those goods and consequently increased the use of coal. According to Jevons, efficiency gains equate to increased, rather than decreased, resource consumption.[10] This somewhat unintuitive economic response was termed "Jevon's Paradox." Today's economists call the idea that efficiency promotes consumption "economic rebound."[11]

Was the improved efficiency in motor vehicle fuel economy under the US CAFE standards wiped out by more use of gasoline? The answer is: "to some extent," and there are two reasons. First was the response of the automobile manufacturers to the imposition of the CAFE standards. One might call it "producer rebound." When car companies found that the fuel economy rules were at odds with their ability to sell passenger vehicles most profitably, they exploited a loophole in the 27.5 mpg rule. There are hard and fast definitions for a passenger car, and the original CAFE standard of 27.5 miles per gallon for this vehicle class has been in force since 1990.[12] However, the definition of what constitutes a light truck is quite lenient, as is the original CAFE mileage standard for this vehicle class, which was below 20.7 mpg through 2004 and increased to 22.5 mpg in 2008. The light truck standard is 5 mpg below the original CAFE standard for cars. The definition of a light truck is a vehicle used primarily for commercial purposes, but the class includes SUVs, minivans, and small pickup trucks. Manufacturers took the liberty of assuming that if the rear seat of a vehicle could be removed, the vehicle met the definition of a light truck subject to the lower fuel economy standard.[13] So vehicles ranging from family vans to the PT Cruiser were considered "light trucks," and with the loophole they were not required to meet the passenger car fuel economy standard. The net result was that the CAFE standards for "cars" were never in full effect.

The second consequence of improved fuel economy under CAFE was indeed a consumer "rebound effect" in the form of increased vehicle use. After the CAFE standards went into effect, Americans began to drive more and opt for less fuel-efficient vehicles. Economists have estimated that for each 10 percent improvement in fuel economy, there is a 1 to 2 percent increase in vehicle travel.[14–16] Total miles driven annually have doubled since 1980, from about 1.5 trillion to 3 trillion miles. This doubling was not simply due to increasing population and a greater number of drivers, as the US population increased by only one-quarter since 1980. In addition, fuel efficiency was not maintained as a top priority of US car buyers. From 1980 to 2004, of all new vehicles sold in the US, light truck sales grew from 16 percent to about 50 percent, with SUV sales growing from 2 to 29 percent. Figure 5.4 shows the sales history of light trucks – defined as SUVs, vans, and small pickups – and new vehicle fuel economy. Because of the increased popularity of bigger and faster vehicles, the US has not fully capitalized on fuel efficiency gains under CAFE. Since the early 1980s, average new vehicle weight climbed to pre-1975 values, horsepower nearly doubled, and 0 to 60 miles per hour acceleration time was cut in half. New vehicles are heavier, more powerful, and faster than before the CAFE standards went into effect.[17] Amazingly, even with these strides toward inefficiency, overall new vehicle fuel economy still met the CAFE standards.

Despite the loss of some of the benefit of increased fuel economy (rebound), gasoline consumption was reduced by efficiency gains. After peaking in 1978, gasoline consumption remained below the peak value for almost 15 years,

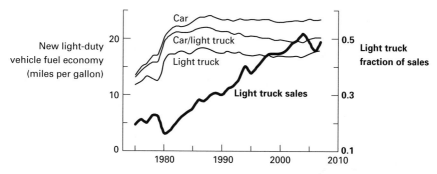

Figure 5.4 Environmental Protection Agency adjusted fuel economy in miles per gallon of new cars and light trucks as well as combined light-duty fleet sales (left axis). Also shown is the fraction of light truck sales (right axis), which now accounts for over half of all new light-duty vehicles. (Data: EPA (2007)[18])

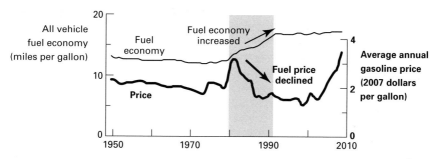

Figure 5.5 Average US car and truck fuel efficiency (upper curve) and inflation-adjusted gasoline price (lower curve). Through the 1980s, fuel economy increased while gasoline prices declined, and yet gasoline consumption fell during most of that period. (Data: fuel economy, EPA; gasoline price, EIA)

until 1993. One might guess that lower consumption resulted from high gasoline prices. Perhaps price was a major influence for a year or two in the US, and during the early 1970s in Europe where fuel-efficient cars were a response to their high-priced gasoline. However, from 1980 to 1993, the inflation-adjusted price of US gasoline declined by half, and gasoline consumption still fell. Efficiency gains did not promote greater consumption during this period. Only after improvement in motor fuel economy began to plateau in the early 1990s (Figure 5.5) did gasoline consumption begin to exceed historical levels (Figure 5.6).

Even with improved motor vehicle fuel efficiency, gasoline use has risen steadily since 1983 and has exceeded the 1980 crest value since 1993 (Figure

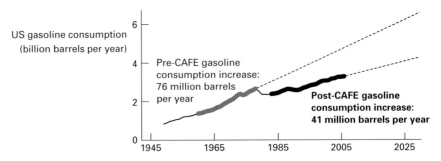

Figure 5.6   The pre-CAFE increasing trend in US gasoline consumption declined from 76 million to 41 million barrels per year after the CAFE standards had their major impact. (Data: EIA)

5.6). So what has ultimately been gained by the increase in motor fuel efficiency required by the CAFE standards? The answer is that the CAFE standards were largely responsible for a structural change in consumption. The rate of increase in fuel consumption never returned to that of the pre-CAFE days. US per capita gasoline consumption fell from 11.9 barrels per year in 1978 to 11 barrels per year today. Had fuel use followed the pre-CAFE trend, 2007 gasoline consumption would have been almost 50 percent higher than actual gasoline use (Figure 5.6), a 1.5 billion barrels of gasoline per year difference. The rate of increase in total US gasoline consumption slowed dramatically in the post-CAFE standards period. Comparing the period since 1984 with the period from 1960 to 1978 (before CAFE was fully implemented), the annual rate of increase in gasoline consumption dropped from 76 to 41 million barrels per year. With less consumption of gasoline, there has also been less consumption of oil. US gasoline consumption is directly related to US oil consumption; as a general guide, the ratio of US gasoline-to-oil use is 0.43. Annual oil consumption is much lower today than it would have been in the absence of CAFE standards.

Much of the reduction in US gasoline consumption was due to the influx of fuel-efficient, foreign-designed vehicles. In 1975, compared to the average US vehicle, the average foreign car was 40 percent lighter and got 50 percent better fuel economy. Foreign-designed cars were 18 percent of the US market in 1975 but accounted for 29 percent of the market in 1980. In 1975, small cars accounted for about half of domestic automobile company sales versus over 95 percent of the sales of imports from Asia and Europe. The post-CAFE, higher fuel-economy era ushered in small foreign cars getting more than 30 mpg and exceeding the CAFE standards.

   US car companies were unprepared to meet consumer demand for fuel-efficient cars, and this translated to an opportunity for manufacturers in Asia and Europe.[19] From 1975 through the 1990s, US manufacturers barely met the CAFE standards while foreign manufacturers surpassed them. Over time, foreign companies gained further market share. They more than doubled the number of cars they produced in the US from 1990 to 2000. Jumping ahead to today, comparing US with foreign manufacturers, there is little difference in the fuel economy of passenger cars they sell.[20] However, by misreading consumer demand for fuel-efficient vehicles, US manufacturers lost their strong foothold to foreign manufacturers. This is a major reason why US car companies have suffered in recent years.

   The increase in fuel efficiency was a global phenomenon. The pre-CAFE years were characterized by a year-over-year increases in global oil production of 0.8 billion barrels, while the post-CAFE years have seen an increase of only 0.3 billion barrels (Figure 5.7), even with rapid growth in developing nations like China. Compared to pre-CAFE oil production, extrapolation of production data suggests worldwide annual oil savings of about 70 percent in 2007, or 19 billion barrels per year. The key point is that to meet the need for transportation fuels, the world produces 27 billion barrels of oil per year versus an estimated 45 billion barrels based on extrapolation of the pre-CAFE period trend.

   Global oil production was lower than the 1979 peak for 15 years in the post-CAFE standards era, even though the per-barrel price of oil was significantly lower after the price crested in 1980 at $93 (2007$). Given low oil prices in the 1980s and the 1990s, averaging $27 (2007$) per barrel, one would have expected oil consumption, and the production required to meet that consumption, to have increased rather than decreased. On the contrary, fuel efficiency gains became fixed in the global oil economy.

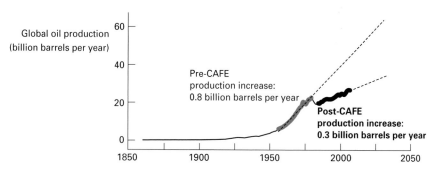

Figure 5.7   Global oil production showing reduction in the trend in the pre- and post-CAFE standards era. (Data: EIA)

The efficiency mandated by the 1975 CAFE standards was accompanied by a 20 percent reduction in global per capita oil production. Global per capita oil production peaked in 1978 at 5.1 barrels per year, and after a rapid four-year decline, the value has stabilized at 4.1 barrels per year since the early 1980s (Figure 5.8). Although total oil consumption was not permanently reduced by greater efficiency, the fact that global *per capita* oil production has remained level for 25 years means that the increase in fuel use tracks the increase in worldwide population. Greater oil production has resulted from the fact that there are more consumers, but on a global basis, efficiency gains have offset increases in oil production needed to meet demand beyond that.

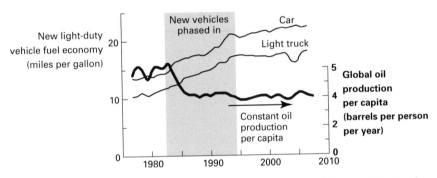

Figure 5.8 Average fuel economy of cars and trucks on the road improved during the years following implementation of the CAFE standards. Simultaneously, per capita global oil production fell from about five to four barrels per year. (Data: fuel economy, EPA; oil production, EIA; population, Economic Research Service, USDA)

The US trend in motor vehicle use is not a sustainable model for the rest of the world. As shown in Figure 5.9, the number of vehicles per capita in the US has doubled since 1950 and now exceeds 840 per 1,000 people. This proportion is 30 times that of China. In terms of vehicle ownership, China is where the US was in 1915. So is Africa. India, with fewer than 20 vehicles per 1,000 people, is even further behind.[21] If the China of today had followed in America's footsteps, it would be home to 4.5 times the number of vehicles on the road as in the US.

Two scenarios for developing nations

It appears that there will be tremendous demand for motor fuel in developing regions. With car fuel efficiency today grounded in the 30-year-old CAFE mandate, if oil continues to be the source of the world's transportation fuel, we are in the midst of a race between technology that builds efficiency and

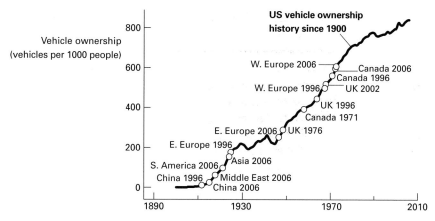

Figure 5.9 Motor vehicles per 1,000 people in the US since 1900 (bold line) compared with vehicle ownership in specific years (1971, 1996, 2002, 2006) in other countries and regions (circles).[22] (Data: DOE and IMF)

the counterbalancing global pressure of increasing motor vehicle use with rising population. Given the demand for oil in new industrial nations, one can imagine two scenarios for their development.

Under one scenario, China and India will not rapidly climb the motor vehicle ownership trajectory laid out by the US and followed, to some degree, by other Western nations. There are three reasons why lower vehicle ownership might be expected in developing nations. The first reason is that car ownership in China, and presumably in India as well, is expected to saturate at about 300 vehicles per 1,000 people (versus over 840 vehicles per 1,000 people in the US). This saturation value is based on a study of the historical increase in vehicle ownership with income growth in Japan, South Korea, Taiwan, and China.[23] Most people in developing countries will not own cars for many years. In 2008, average income in China was approximately $3,200, and in India it was $1,100.[24] These values are a fraction of income in the US ($47,000) and European Union (about $38,500). In China, gasoline is only slightly less expensive than in the US, and yet Chinese incomes are much lower. Annual consumption of 600 gallons of gasoline in the US costs consumers $1,200 to $2,400 per year at $2 and $4 per gallon, respectively, and represents less than 5 percent of average income. In China, 600 gallons would cost $1,100 to $2,200 per year, or an amount averaging half the entire Chinese average income. In India, the largest car manufacturer is Tata Motors, which in 2008 introduced the 50-mpg Nano at a price of $2,000.[25] Although this car perhaps could be purchased by many in India, the costs to run and maintain a car put it beyond reach of the average Indian family. Even if Indian and Chinese incomes doubled every seven years (10 percent growth), which does not seem likely given the world economy after the 2008 global financial crisis, cars purchased at any price would remain too high to operate for most people in China and India for some years to come.

The second limit on growth in vehicle ownership, or at least the reduction in vehicle miles traveled, is degradation of local air quality from so many millions of vehicles in dense industrial urban areas.[26] Local and regional air quality protection may turn out to be the major factor that keeps transportation fuel and automotive use in check. The third limitation on vehicle ownership growth in industrializing nations is that greater oil demand, sparking high fuel prices like those in 2008, would stifle economic growth. The high price of oil and gasoline was painful to consumers in the US and Europe, but imagine how such a price might suffocate industrial growth in the developing world. Such a high price of oil could impact China by driving up manufacturing transportation costs, which would in turn reduce the attractiveness of their exports. Combined, these two factors would tend to slow Chinese industrial growth and put additional pressure on Chinese consumers faced with high transportation fuel costs. However, if this chain of events occurs, ultimately, lower demand for oil by China should tend to force down global oil prices.

A second development scenario is that people in industrializing nations will rapidly own greater numbers of motor vehicles and increase their oil consumption in a repeat of what occurred in the developed world. Under this growth scenario, the world seems headed toward a fuel crisis generated by a transportation-demand impasse – over one billion cars in China alone! It wouldn't be the first time that such an unsustainable situation appeared imminent. Under this scenario, a drastic change in transportation must occur, as it has in the past.

Consider the period before automobiles. In 1900, there were 130,000 horses in New York City. These horses were used almost entirely to transport goods, not people. With only 1,500 horse-drawn taxis, urban dwellers didn't rely on horses to get around, and certainly did not ride to work on horseback. As it was, in Manhattan there was already one horse for every 26 people. The problems with horses were serious. First, there were 5.7 annual "traffic" fatalities per 100,000 people from accidents involving wagons and carriages. Current New York City traffic fatalities are only 3.7 per 100,000 people per year, and since 1900 the urban population has grown from 3.4 million to 8 million.[27] Second, 130,000 horses produce about 1.5 billion pounds of manure per year (enough to cover an acre to a depth of about 550 feet). Third, as early as the 1880s, 15,000 dead horses had to be removed from the city every year. Similar problems with horses existed in St. Louis, Chicago, Philadelphia, and Boston, which had between 32,000 and 74,000 horses each, and many thousands of horses occupied the streets of smaller cities like San Francisco.[28]

Urban horses and the massive problems of sanitation and safety they generated were replaced by cars in the early twentieth century. In 1900, there were only 8,000 cars in the US. That number climbed to 458,000 in 1910 and then to 8.1 million in 1920. Car owners were primarily found in cities. New York City had 2,400 cars in 1900 and by 1920 it had 213,000, a greater number than the historical horse population. In St. Louis, there were no cars in 1900. By 1914 there were over 10,000, and by 1920, about 48,000. During that period, the St. Louis horse population declined from a peak of 35,000 in 1908 to fewer than 20,000 in 1920. Cars phased in and horses phased out.

Compared with the problem of too many horses one hundred years ago, the problem of too many cars does not seem so terrible. Returning to the second scenario for developing nations, the current upward trajectory of car ownership and fuel consumption is not sustainable. Rather, a sudden shift in technology is needed to change the landscape. While the world was caught up in managing horses, substitution of cars for horses eliminated a host of problems in the course of a few decades, or at least traded those problems in for new ones. Perhaps such a sea change is on the horizon for cars.

What Might Ultimately Substitute for Oil?

The history of natural resource use is full of examples of substitution. As a resource becomes expensive, substitutes emerge. These substitutes may be expensive initially, but over time they replace the uses of the original resource. Should oil prices return to high levels like those of 2008 (for example, oil supply becomes unreliable due to world events), transportation fuel substitutes made from abundant coal, natural gas, and oil shale resources would become competitive. Fuel options expand. The significance of the distinction between conventional and unconventional oil diminishes, just as with oil sands and heavy oil already. If the production cost, including safety and environmental impacts, of a fuel suits transportation needs and is competitive with oil, it doesn't matter whether that fuel is from refined oil or derived from baked oil shale. Yet, in addition to fuel cost, there are other factors influencing the global selection of substitutes for transportation fuels. These factors include dependence on oil imports, atmospheric pollution, and the rate of technological innovation that generates alternatives. Let's consider these factors.

Consideration 1: Cost of dependence on imported oil

Since the early 1970s, the US has relied heavily on imported oil. The economic impacts over time of importing oil have been measured in two ways.[29,30] The first measure is the cost of oil imports relative to gross domestic product (GDP), as shown in Figure 5.10. By this measure, the US is seemingly approaching a level of dependence not seen since the oil crisis of 1979, when gasoline supply was so limited that there were long lines of cars at gas stations. The economic recession of 1980 to 1982 followed. Even though the US imported over 50 percent more oil from 2005 to 2008 than it did from 1977 to 1980, there was similar dependence on OPEC, which satisfied nearly 30 percent of US crude oil demand in both periods. The second measure of economic impact is to evaluate the decline in GDP due to oil imports. The US economy is so dependent on oil that the added expense of imported oil, which ends up in the pockets of exporters, has negative ripple effects on the US GDP. These effects are complicated, and to compute GDP losses over time due to oil imports requires many assumptions, but economists David Greene and Paul Leiby at Oak Ridge National Laboratory have developed a mathematical model to do it. Their estimates of annual GDP loss are shown in Figure 5.10. The curve of GDP losses resembles the curve for the first measure of economic impact – the history of oil import costs relative to GDP. The peak percent GDP loss occurred in 1982, near the end of the US recession. In 2007, the US lost an estimated 1.7 percent of its GDP as a consequence of importing oil. This loss reduced 2007 GDP growth by more than half. However, the effect of oil imports on GDP was more pronounced in 1982 because GDP growth during that recession was negative (minus 2 percent), and the estimated reduction in GDP was higher at 3 percent. So 2007 and 1982 were quite different. But the world economy has changed dramatically since 2007, and we might benefit by looking back once more to 1982.

Lower oil prices during the economic crisis of 2008–9 occurred because of the sudden reduction in global oil demand. As the world economy entered into recession, the impact of US dependence on oil imports was mitigated by a decline in the price of oil and other commodities. Each $1 per gallon decline in the annual average gasoline price (from $3.25 in 2008) is akin to an annual tax cut of $138 billion. However, the US has suffered its worst economy since the 1930s. If there is a period like the early 1980s, when the price of gasoline was $3.12 per gallon (2007$) and oil reached $93 per barrel (2007$), $100 per barrel oil could spur a 3 percent reduction in US GDP resulting from oil imports. Under this scenario, oil importation would compound a decline in US GDP. The US economy in recession benefitted from the oil price decline of 2008–9 but remains vulnerable to an oil-price shock, potentially being hit while it is down.

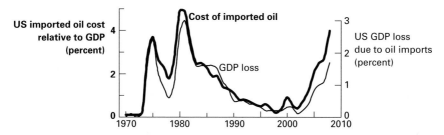

Figure 5.10   The cost of oil dependence, showing the cost of oil imports relative to US gross domestic product (GDP) and the consequent losses in GDP. (From: Greene (2008), "Oil Security Metrics Model"[31])

Consideration 2: Gasoline and atmospheric carbon dioxide emissions

A second major factor that comes into play when weighing alternatives to oil is air pollution, and more specifically, the influence of carbon dioxide emissions on global warming. Carbon dioxide, an unavoidable by-product of combustion, is a greenhouse gas. Carbon dioxide allows light, ultraviolet rays, and other radiation from the sun to penetrate the atmosphere, but it absorbs the heat energy that re-radiates back from Earth's surface. Atmospheric carbon dioxide levels have increased markedly in the past 100 years, tracking the global consumption of fossil fuels that has accompanied increasing industrialization.

Burning fossil fuels is the major source of increased levels of atmospheric carbon dioxide. Worldwide, oil use is responsible for 39 percent of carbon dioxide emissions. One-third of US carbon dioxide emissions are from transportation fuels, and the majority (60 percent) of that quantity stems from gasoline use.[32] In some ways, the emission of carbon dioxide from the combustion of fossil fuels has motivated technological change just as the pollution from urban horses did 100 years ago.

US transportation generates more carbon dioxide pollution than either the industrial or residential sectors, which depend on all types of fossil fuels – oil, natural gas, and coal. While China is number 1 in carbon dioxide pollution, mainly from burning coal, the US emits twice as much as any other country from oil use. In 2007, the typical American car owner used about 550 gallons of gasoline and added 11,000 pounds of carbon dioxide to the atmosphere, or about one pound per mile. Every gallon of gasoline consumed generates 20 pounds of carbon dioxide.[33] The US was responsible for emitting about 6 billion metric tons of carbon dioxide to the atmosphere (Figure 5.11), or over 20 percent of the global total (29 billion metric tons). US gasoline consumption alone emits 1.2 billion metric tons per year, and when all fuels from oil are considered, the value is about 2.6 billion metric tons. To get a physical feel for carbon emissions, consider that the weight of carbon dioxide emitted into the atmosphere from gasoline consumed in the US is more than twice the weight of all US cars and light trucks – and that emission occurs in just one year.

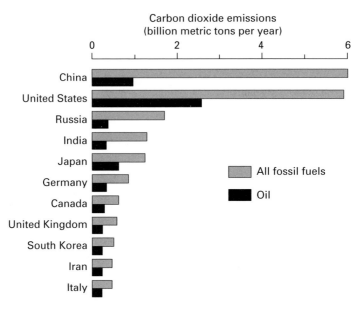

Figure 5.11 Carbon dioxide emitted to the atmosphere from all fossil fuel and oil combustion, respectively, in 2006. Global emissions are 29.2 and 11.2 billion metric tons from fossil fuels and oil, respectively. The 11 countries shown account for two-thirds of global emissions from fossil fuels, with China and the US accounting for 40 percent. (Data: EIA)

Consideration 3: Alternatives

The benefit of potentially sustained high oil prices is that new products emerge as substitute fuels. As discussed in Chapter 4, liquid transportation fuel can be made from both natural gas and coal. Natural gas, which is among the cleanest-burning fuels, is already used to power motor vehicles. In addition, biofuels are a potential alternative, but they are not a panacea.

Ethanol

Ethanol is the alcohol in wine, beer, and liquor. As a biofuel, it is produced from the fermentation of various types of common cultivated crops. Ethanol from corn in the US and sugar beets in Europe has become a widespread component of gasoline, but as a pure transportation-fuel alternative, its use presents several problems. First, compared to gasoline, it costs about $1 per gallon more to produce, in part because of the competing demand for corn as food for livestock, and the cost of corn accounts for three-quarters of the cost of ethanol production. Second, by depending on corn as a transportation fuel, weather and climate can negatively affect both food production and transportation.

Ethanol can also be made from a variety of plant stocks. For example, Brazil produces a quarter of its transportation fuel from ethanol derived from sugar cane, which can be converted to ethanol more efficiently than can corn. In addition, there are hopes that ethanol can be produced on a large scale from plant materials such as wood chips and the stalks and leaves of corn, but a large-scale, commercially viable, cellulosic biomass conversion process has presented some technological obstacles.

Yet, there are still some fundamental problems with refining ethanol from any source. Ethanol has an energy density of merely three-fifths that of gasoline on a per volume basis, so as a fuel, it takes 10 gallons of ethanol to substitute for 6 gallons of gasoline. Ethanol is more corrosive than gasoline, and using ethanol as a transportation fuel would require significant changes in the fuel-supply infrastructure, from pipelines to storage tanks.[34] Finally, most cars are not designed to run on pure ethanol. Gasohol is a fuel with the highest blend that today's cars can handle, and it contains just 15 percent ethanol.

One effect of the world moving toward ethanol produced from corn and sugar is a shift in land use. One might think that such a land-use change would reduce global carbon dioxide emissions because new plant material would sequester carbon, rather than emitting it when gasoline is burned. In fact, some studies have suggested that burning corn ethanol would reduce greenhouse gas emissions by 20 percent as compared with burning gasoline. However, a 2009 study concluded that if forest and grassland are converted to corn fields, greenhouse gas emissions would double over a 30-year period compared to preserving existing land uses and continuing to burn gasoline. Carbon sequestration is significantly diminished when existing forest and grassland is changed to cropland.[35] In addition, another analysis suggests that the costs of using corn ethanol, in terms of greenhouse gases and particulate emissions (with negative health effects), may be up to twice those of using gasoline, depending upon the heat source used to refine the ethanol.[36]

By 2009, ethanol had already gone through a boom and bust cycle in the US. First came the boom. In 2008, one-third of corn grown in the US was used to make ethanol at over 150 processing plants. This represented 6.7 percent of US gasoline supply (138 billion gallons per year), or 4 percent of gasoline supply in energy-equivalent terms. US ethanol production of 9.2 billion gallons in 2008 exceeded the 2005 Energy Policy Act mandate to produce 7.5 billion gallons of renewable fuel by 2012. Given the federal subsidy of 51 cents per gallon of ethanol provided to gasoline blenders, the US Department of Agriculture projected that by around 2015, annual ethanol production would increase to 12 billion gallons.[37–40] Then came the bust in early 2009. With the sudden drop in oil prices and the decline in demand for transportation fuel, the ethanol business hit a wall. The regulatory limit to the amount of ethanol that could be blended with gasoline was set at 10 percent, and as gasoline consumption declined, so did the need for corn ethanol. By early 2009, 24 ethanol plants had shut down.[41]

There are serious doubts about the future of corn ethanol as a viable, large-scale, substitute transportation fuel. Of the 325 million acres of cropland in the US, about 30 million were used to produce corn ethanol. If all of the croplands in the US were used to produce ethanol from corn, it would not satisfy the equivalent US gasoline needs. That is, the 100 billion gallons of ethanol made from a blanket of corn planted in the US would be less than the 138 billion gallons of gasoline consumed in 2008. In energy equivalent terms (60 billion barrels of gasoline equivalent), this ethanol would displace less than half of 2008 US gasoline demand.

Biodiesel

Rudolf Diesel demonstrated his new engine at the 1900 World Fair using peanut oil for fuel. As it turned out, petroleum was found to be a less expensive stock for diesel fuel. Today, however, biodiesel has re-emerged as an alternative to diesel produced from petroleum. Biodiesel fuel is made by a simple chemical process from any one of a variety of vegetable oils, such as soybean and palm, as well as animal fats from waste cooking grease. Biodiesel has some beneficial characteristics as a fuel. One is that it readily autoignites, even more so than petroleum diesel. Another characteristic is good lubrication properties, unlike low- or even ultra-low petroleum diesel.

On the other hand, biodiesel has its problems. It has a 5 percent lower energy density (per volume) than petroleum diesel and has a "drivability problem": at low temperatures, biodiesel is more likely than petroleum diesel to clog engines and filters with wax crystals. In addition, as a global fuel, biodiesel presents several environmental issues. For example, palm oil has been produced as a source of biodiesel in Indonesia but at the cost of the destruction of rainforests and loss of orangutan habitat.[42] Some potential oil stocks are limited by competition from other industries. For example, in the US, it is anticipated that production of biodiesel will be limited because of the cost of soybean oil, which is needed by both the food and soap industries. Cooking oil, another potential biodiesel stock, has limited availability.

Leapfrogging to an ultimate substitute

Two characteristics of the historical production of other Earth resources might be supposed to enter into the oil-substitution landscape: meeting **end-use** needs and **technology leapfrogging**. Does the world need oil? Although the modern world seems to run on oil, for the most part, what is needed is trans-portation, not fuel from oil. The movement of people and things is the real requirement. Oil is "simply" a convenience. However, if our future world does not depend on oil, how will we meet the end-use that it now serves? Perhaps the need for transportation will not be met directly by any liquid or gas. After all, these are just readily dispensed substances that temporarily store and release energy.

Horses and liquid fuels are not the only options that enable transportation. It is actually difficult to imagine that cars of the future will burn gasoline or diesel fuel, no matter what their origin. If the means of propulsion of motor vehicles leapfrogs over substitute combustible liquids, it will likely land on electricity. Near the turn of the last century, electricity, not natural gas or kerosene, became the ultimate source for production of nighttime lighting. Electricity, not oil, has provided energy to industry.

Consider the Electric Vehicle Company, which was seeded with $20 million in securities. Its mission was to develop electric cars for use in urban areas. The company produced about 2,000 electric taxis that ran on the streets of New York, Chicago, Philadelphia, Boston, and Washington. But both the battery technology and business schemes the company promoted did not prove out, and the effort failed. The prospects for a non-polluting car were dashed. What is interesting is that the Electric Vehicle Company was founded in the year 1897 and went into receivership in 1907.[43] There have been peri-odic incarnations of electric cars over the past century. The question is, has the day of the electric car finally arrived?

For transportation, there are advantages of electricity over gasoline. Elec-tricity can be produced from a variety of sources. Any fuel can be used to generate it, and it can be derived from hydropower and nuclear energy, from solar and wind energy, and from geothermal and biofuel sources. Using electricity for transportation could reduce dependence on foreign oil. Because electricity cannot be readily stored, generating facilities must operate continuously, despite fluctuating demand. Even though demand is much lower at night, power generation continues, resulting in excess electricity being available at night; much of this electricity goes unused. Therefore, power plants represent an untapped resource of energy that could be used to power transportation with no increase in generating costs during off-peak periods.

Electricity is advantageous because it can be produced on a variety of scales and can be less polluting than cars and trucks that burn gasoline. At the largest scale, centralized electricity generation potentially could capture carbon before it is emitted to the atmosphere, or it could be based on non-combustion technology that emits little carbon. Capturing carbon emitted from a centralized power plant is easier than trying to stop carbon emissions from hundreds of millions of individual motor vehicles burning liquid fuels. At the smallest scale, decentralized solar power generation would avoid direct carbon generation.

New technology has improved the possibility of large-scale production of electric cars, but significant hurdles must be overcome. One attraction of a rechargeable electric car is that plugging in at your home would cost 25 to 40 percent of the cost of gasoline.[44] Rechargeable lithium-ion batteries have been used successfully in an electric vehicle that can go up to 220 miles on a single charge. Battery technology has improved but there are still issues of vehicle range, battery weight, battery life, recharge duration, and impacts on regional power systems. Made by Tesla Motors, Inc. in California, one high-performance electric sports car, which sells for $110,000, can travel over 150 miles on a single-night's charge.[45] The company has unveiled a sedan starting at $56,000 with an expected range of 160 to 300 miles per charge, depending on the selected battery option.[46] Daimler plans to use Tesla's battery pack in its future electric Smart car, which is bound to be more moderately priced.[47] Companies touting electric vehicles have generated enthusiasm over the years, but the industry remains limited by a viable battery technology and business model (i.e., producing cars at a price that masses of drivers can afford) – issues reminiscent of the Electric Vehicle Company of 1897.

The virtue of battery-powered electric cars is that electric motors are very efficient. In fact, they are about three times more efficient than internal combustion engines in equivalent small SUVs for highway driving. Electricity is the ultimate flex-fuel and can be produced by combustion of a variety of fuel-stocks. It is estimated that an acre of switchgrass (a fast-growing, perennial, tall prairie grass) used as a fuel to co-produce electricity at power plants could increase travel distance by 56 percent compared with converting that same cellulosic biomass to ethanol and using it directly as a transportation fuel.[48] Technological advances in battery technology will be key to ushering in the era of the electric car. One such breakthrough was reported in March 2009 in the journal *Nature*. Researchers at MIT developed a small battery that can be charged in 10 to 20 seconds rather than the six minutes that it traditionally takes. Their new lithium-ion battery allows the lithium ions to move rapidly through a newly developed material with an ion-conducting surface. Furthermore, compared to current battery materials, the new battery material does not degrade as much on recharging.[49] If this new technology is proven to scale up to electric-car batteries, an eight-hour charging period could be reduced to about 30 minutes.

Some view plug-in hybrid electric vehicles, which rely on both liquid fuel and battery power, as a way to reduce gasoline consumption and provide a potential bridge to a world with fully electric vehicles.[50,51] Plans are for plug-in hybrids to be produced by companies like Honda (Prius), BMW (Mini E), Chevy (Volt), and BYD, which stands for Build Your Dreams (F3DM), a Chinese company with a 10 percent stake owned by investor Warren Buffett.[52] Plug-in hybrids, even with a very limited driving range per charge, are so efficient that they would significantly reduce carbon emissions. Plug-in hybrid cars that could travel just 10 to 40 miles per charge of electricity derived from a totally coal-based US power grid would produce 25 to 30 percent lower carbon emissions than cars with conventional engines.[53–55]

Considering the regional implications of charging the batteries of electric cars, one study evaluated the impact on US power plants of plug-in hybrid-electric vehicles.[56] The study assumed 25 percent market share of plug-in electric passenger cars by 2020. The number of vehicles and energy demand were assessed by US regions defined by the EIA. It would take four to eight hours to recharge a car's battery, depending on the vehicle type and assuming only local travel. Because there is excess power plant capacity during off-peak hours, recharging the batteries of tens of millions of plug-in cars is assumed to be restricted to evening and nighttime hours. Even with all vehicle batteries being charged at night, new power plants would be required to meet the peak capacity energy demand. The increase in annual peak capacity energy demand due to plug-in electric hybrids is projected to be as high as 28 percent (in the California region) by 2030. Although there are serious technological and infrastructural challenges for the production of electric cars, the electric hybrid offers a valuable option to transition away from gasoline and diesel fuel for transportation.

Effects of a US move to oil alternatives

What are the implications of the US moving toward energy independence by lowering its reliance on oil as a transportation fuel? At face value, reducing the use of oil seems to have obvious benefits. First, such a move would be accompanied by the sustainable development of renewable energy resources including solar, wind, and wave energy, and perhaps biofuels. Second, moving away from oil would reduce carbon emissions. Third, an oil-free economy would be less susceptible to disruptions in imported supply and consequent price shocks.

However, it is not evident that the presumed benefits listed above would materialize. If the US were to rapidly and dramatically reduce its use of oil, global oil demand would decline, and the excess in global supply would push down the price of oil. Developing nations would take advantage of the lower-priced oil to build industry and grow their economies. In essence, if the US rapidly switched from oil to alternative energy sources, much of the economic benefit would be absorbed by rapidly growing developing nations. The growth of developing nations helped by inexpensive oil would not be an economic sacrifice to the US if the cost of using alternative fuels were competitive with the cost of oil. Unfortunately, the direct cost of current alternatives has been too high.

What about reducing carbon emissions? The emission of carbon dioxide into the atmosphere is a global problem that is immune to where the carbon is released. Suppose the US reduced its oil consumption and its national carbon footprint. Once again, if oil that is no longer demanded by the US is consumed by developing nations, their loading of carbon to the atmosphere would grow as America's declined. To reduce global carbon emissions, all nations must participate; merely shifting emissions from one place to another would have no global benefit. It does not appear that developing nations like China and India will sacrifice economic growth to support a global community effort to move from oil to alternatives merely to produce less carbon dioxide. Indeed, by 2015 China plans to increase its coal production by 30 percent.[57]

Given current power plants and proven technology, shifting to an electricity-based transportation fleet that depends on burning coal in power plants could increase rather than decrease global carbon dioxide emissions. No commercial power plants remove carbon dioxide, and coal-fired power plants are responsible for 40 percent of global carbon dioxide emissions.[58,59] Much is made of **"clean" coal technology**, but there has not been a commercial demonstration of a viable technology that removes carbon dioxide from power-plant coal combustion.[60] Emissions from coal are cleansed of impurities like sulfur, nitrogen oxides, mercury, and particulate matter, but not carbon dioxide. It is possible to collect the carbon dioxide produced by power plants and sequester it by injecting this greenhouse gas deep underground. However, the feasibility of successfully deploying the technology is controversial and unproven at the large scale at power-plant locations where the carbon dioxide is produced. The US DOE has set the goal of promoting the development of coal-based power generation systems that remove 90 percent of the carbon dioxide for an electricity cost increment of less than 10 percent. Such technology is at best 15 years away.[61] The problem is that new coal-fired power plants that emit carbon dioxide are being built every week in China and India.[62] Together, China and India are already responsible for one-third of global carbon dioxide emissions from their power plants and industrial production. It is estimated that 79 percent of the increase in global coal use by 2030 will come from China and India.[63] In the US, five new coal-fired power plants went into operation in 2008, and another 28 were under construction.[64] In Europe, 50 new coal-fired power plants are planned to come on line by 2014.[65] All new plants are built to last for 40 years.

The US might indeed be protected from economic oil-shocks if transportation no longer depended on oil imports. But what would happen to countries whose economies depend on oil exports? Without revenues from oil exports, some nations would likely face political, economic, and social turmoil. The consulting company PricewaterhouseCoopers classifies the political stability of nations based on the state of their government, society, security, and economy.[66] Consider seven well-known oil-exporting countries: Iran, Iraq, Nigeria, Libya, Venezuela, Ecuador, and Indonesia. All seven countries have political stability rankings of either low or very low, which places them at the bottom of the six-category classification. Furthermore, Transparency International ranks the perceived corruption of 180 nations.[67] Its 2008 rankings go from a least corrupt score of 9.3 (Denmark, New Zealand, Sweden) down to a highly corrupt 1.0 (Somalia). The seven oil-exporting countries listed above have an average corruption index of 2.2, with Nigeria having the best ranking of the group at 2.7. The relative political instability and lawlessness of these countries is perhaps an indication that they could not tolerate elimination of the significant income they obtain from oil exports. The point is that one potential unintended consequence of a reduction in oil use is social disintegration of some oil-exporting nations.

How do oil price and price stability affect the future of transportation fuels? Consumers like inexpensive gasoline. In the US, the drop in gasoline price in 2009 to its long-term, historical average of $2.25 per gallon was a relief and economic benefit. However, at that price, the development of new, expensive offshore oil sources and alternative liquid fuels is not profitable. So the dilemma is that low prices are welcome but deter the creation of future supplies of transportation fuels. An extended period of low oil prices would also strengthen the hand of the lowest-cost oil producers in the Middle East (i.e., some of the OPEC members).

The problem of "oil priced too low" may resolve itself if OPEC sustains production cuts that force oil prices to increase. Low-priced OPEC oil producers in the Middle East have remained profitable throughout all periods of oil-price volatility. The cartel made significant strides toward achieving its target of $75 per barrel (at least temporarily) when the price of oil broke the $70-per-barrel level in June 2009. The trouble is that OPEC's fair price target might not stop at $75 per barrel, and there are few ways that the non-OPEC world can combat OPEC's production-based price controls. One way in the long run is a switch from oil to alternatives. But further evolution of economically viable fuel alternatives and fuel-efficient vehicles requires time.

In the short run, it would seem that only a cartel that competes with OPEC could guarantee oil price stability. In principle, the US could store oil in an "economic petroleum reserve" and provide supply stability if OPEC cut production to push the price of oil to $200 per barrel. The OPEC oil embargo of 1973–4 revealed the vulnerability of the US to an oil-supply shock. Consequently, beginning in 1977, the US established the **Strategic Petroleum Reserve (SPR)**, which can be used at the discretion of the President in the event of an emergency. Oil is stored in huge manmade subsurface caverns in salt deposits about 3,000 feet deep along the US coast of the Gulf of Mexico. The SPR maintains only enough oil to offset about two months of imported oil. The SPR has been tapped only twice for emergencies: once in 1991 during the first US war in the Persian Gulf and again in 2005 after Hurricane Katrina.[68] Plans to more than double the SPR have been criticized, as the existing stockpile has been mismanaged.[69]

If the US had the capacity to store 3.8 billion of barrels of oil apart from the SPR, it would be enough to offset over a year of imported oil, or almost two years of imports from OPEC.[70] Much less capacity would be required if finished fuel products were stored. Such a national oil bank could provide a backstop to future OPEC production cuts aimed at raising the price of oil. The benefits of storing this amount of oil would be the US's ability to: (1) stabilize and moderate the price of oil, (2) temper OPEC's cartel power, (3) provide the US with a significant buffer for transportation energy security, (4) serve as a valuable and appreciating asset in the event of rampant inflation, and (5) allow an extended and stable transition period for the development of alternative transportation energy resources.

Of course, the premise is that low-priced oil would be used to fill an "economic petroleum reserve," and doing that would take time and skillful management. In the short term, oil exporters would benefit, since additional oil would be needed to fill the stockpile. In fact, most of the oil used to initially fill the SPR in 1977 was Saudi Arabian light crude. In the long run, the wise use of an oil stockpile would provide greater price stability and greater security. The power of the OPEC cartel could be significantly tempered if most major oil-importing nations also built their own oil stockpiles, creating a *de facto* cartel consisting of non-OPEC members (perhaps called *NOPEC*). The US, as the consumer of about one-quarter of the annual global oil supply, is in a strong position to take the lead in countering OPEC reductions in exports and maintaining oil-price stability.

Developing a US "economic petroleum reserve," or national oil bank, would require 5.5 times the amount of oil stored in the SPR and an investment of perhaps $130 to $230 billion in oil purchases. The other option for the US and other oil-importing nations is to trust that: (1) OPEC will continue to provide oil and no longer manipulate price through production limitations, (2) non-OPEC suppliers will discover and develop new sources of oil even if it is not economic for them to do so, (3) violent oil-price swings will not severely disrupt global or regional economies, (4) prospective conventional oil substitutes will not be economically stranded by periods of low oil prices following sustained periods of high prices, and (5) cost-effective alternative energy sources will be brought online rapidly.

Given the state of global oil availability, there is time to develop, test, compare, and transition to optimal alternative technologies to power transportation. Panic as a response to oil-price spikes and the perception of scarcity is likely to lead to government promotion of losing technologies, for example, subsidizing corn-based ethanol production. The public policy challenge is to maintain the stability of oil supply and price while converting to environmentally sustainable and cost-effective sources of energy for transportation. We are nearing a crossroad in this regard. The economic crisis that began in 2008 resulted in a reduction in new car purchases. In 2005, 17 million cars and light trucks were sold in the US, but this number had dropped to about 9.5 million by mid-2009 as consumer spending declined.[71] At that sales rate, it would take about 25 years to replace the US light vehicle fleet.[72] When economic conditions improve, an increase in new car sales will provide a great opportunity to introduce energy-efficient vehicles to replace old conventional ones. Furthermore, in 2008 the average US car had been on the road for 9.4 years, up from 5.7 years in 1978.[73] So, consumers have a pent-up demand for new cars, and the ones they purchase will remain on the road for a decade.

The State of Oil Resources

Before briefly concluding our discussion of the global oil resources debate, let us summarize some key points about oil availability, demand, the environment, and security.

Availability

- The amount of exploitable global oil, estimated as reserves or endowment, is at an all-time high and has risen consistently since values were first reported. This is so even though the discovery of new giant oil fields has fallen since the 1970s.
- Unconventional oil resources are vast.
- Natural gas and coal are plentiful. Natural gas can serve as a transportation fuel. Both natural gas and coal can be converted into diesel or electricity, both of which can serve as transportation-fuel substitutes for oil.

Demand

- Global oil use has grown in direct proportion to population.
- Demand for oil continues to rise in developing nations.
- Use of diesel vehicles and fuel has climbed rapidly in Europe, becoming the transportation fuel of choice in the region.

Environment and Efficiency

- Oil is the source of about 40 percent of global emissions of carbon dioxide (a greenhouse gas).
- If reduced oil consumption in developed nations is offset by increased oil consumption in developing nations, there is no benefit in terms of lowering global carbon emissions.
- Control of carbon emissions from vehicles depends on improvements in vehicle fuel efficiency or a transition to a greater use of electricity, both of which would reduce oil demand.
- Oil-use efficiency has increased steadily, as measured by the decline in oil-use intensity. The rate of efficiency gain in China has significantly exceeded the worldwide average rate.
- Automotive fuel efficiency standards created a large and permanent reduction in the rate of oil consumption. More stringent standards represent a rapid means to reduce oil use.
- Development of an electric car for widespread use has been on the horizon for more than a century. The chief impediments have been viable battery technology and cost.
- Plug-in electric vehicles are a logical transition to fully electric cars.

Security and Stability

- Many oil accumulations exist in politically unstable regions.
- Producing 44 percent of the world's oil, OPEC administers significant production-based price control.
- High oil prices materially disrupt regional economies while they encourage new oil discovery and recovery.
- Sustained low oil prices can devastate oil-exporting nations and can economically strand both high-cost oil development and unconventional oil recovery.
- The US, Europe, China, and India depend heavily on imported oil.
- Given such strong import-dependence, disruptions in oil supply are likely.
- The influence of the OPEC cartel could be tempered by creating large national economic oil storage banks in oil-importing countries that counteract OPEC's production-based increases in the price of oil. Price stability would be maintained as transportation-fuel alternatives were developed and tested.

Ending Thoughts

The serious debate about global oil depletion is *not* merely characterized by optimists on one side and pessimists on the other. Informed positions transcend attitudes. Facts and logical analyses have been used to support the case for impending depletion as well as the case for the continued availability of oil. Those who have thought seriously about the issues have adopted key assumptions that have led them to particular conclusions about the future of oil. An accepted engineering approach is to begin with the idea that there is a finite amount of oil, and the lifetime of the oil endowment depends on the rates at which oil is produced over time. Based on this approach, the timing and trajectory of depletion is plainly a matter of estimating the global oil endowment, evaluating production data, adopting a curve that is fit to those data, and then forecasting production. This approach assumes that the global oil endowment is a known quantity and that consumers will take whatever oil they can get, because production is the limiting factor that controls delivery. It is easy to apply a formula to forecast the production trend based on a global oil mass-balance. But the foundation of this logical approach is lost when two assumptions are disproved. First, take away the notion that an accurate mass balance is possible, because resource assessments have generally projected larger and larger estimates of the global oil endowment with time. Second, eliminate the idea that oil use has been limited by production. Most of the time, production does not limit consumption; rather, consumer demand for oil at a particular price dictates how much oil is produced.

Much evidence shows a declining trend in huge oil discoveries, and oil is seemingly becoming harder to find. Yet, the amount of global oil reserves (known and profitable resource) continues to grow, and the success rate of exploratory wells has increased. The debate continues, with some pointing to recent relatively high oil prices and the slow pace of addition to supplies compared to the past. But the low price of oil (less than $30 per barrel (2007$)) during the 1980s and 1990s discouraged exploration, a trend that reversed when oil prices rose in the first decade of this millennium. All the while, OPEC has tried to control its production and allow oil prices to climb as demand increased, while world events, ranging from striking workers to hurricanes, limited global production. Under these constraints of exploration and production, a snapshot of the high price of oil at a particular time is not a valid indicator of resource scarcity. This is most apparent in the long-term downward trend in the price of gasoline adjusted for inflation.

Compared with the mass-balance and curve-fitting approach discussed above, is it not as scientific to develop a forecast of global oil production that: (1) is based on an unknown oil endowment, (2) allows for oil resources from unconventional sources, (3) considers the dynamics of demand for oil and the purposes oil serves, and (4) anticipates technology that will enhance discovery, recovery, use-efficiency, and the role of substitutes? Let us examine this approach.

Production trends of non-renewable Earth resources have depended on new technologies for discovery, resource development, and efficient use. The difficulty with depending on future technology to make resources available is the risk that advances will not be timely or not occur at all. After all, how can one predict the rate and significance of commercially viable breakthroughs in resource discovery, production, and use-efficiency? One approach is to estimate the historical number of innovations based on inspection of spending on research and development. Another approach is to count patent activity. Both measures can provide some idea of the rate of innovation but are deficient because not all spending or ideas result in successful inventions in the form of new machines or methods, and the lag time between concept and implementation can be decades. Studies of technology in the petroleum industry have attempted to overcome these weaknesses by directly estimating the number of innovative technologies historically brought to practice. The advantage of this approach is the ability to exclude the number of failures from the total and record the implementation date. In two studies, researchers combed through trade journals and consulted with industry experts regarding advances in petroleum exploration and development from 1947 to 1965 and 1966 to 1990, respectively. Both studies found an average of four to five such innovations per year.[74] What these results suggest is that innovations are not rare events. Predicting the likely impact of any new technology remains a challenge, but there appears to be a steady stream of inventions.

Petroleum exploration, development, and refining processes have always been driven by technological advances. From the beginning of the Oil Era, new ways have been used to deduce the presence of oil beneath the land or ocean and to advance the ability to drill, optimally produce oil, and more effectively refine it. In 1908, Howard Hughes, Senior, invented the "rock bit," whose 166 cutting edges enabled drilling deep wells for the first time.[75] In 1913, a patent was issued on the process of distilling oil under pressure wherein heavier hydrocarbon molecules were cracked (broken down) to form lighter liquids. This breakthrough resulted in refining that doubled the amount of gasoline produced from a barrel of oil (40 percent versus 20 percent gasoline).[76]

Today, the oil industry is a high-tech business, with technological advances being adopted in areas ranging from discovery to recovery. Computer-aided visualization, 3D viewing, and new geophysical methods have accompanied major progress in the use of horizontal wells and precision directional drilling. Rather than leaving behind most of the oil after pumping, modern production can remove the majority of oil initially in place. Dismissing the consistent role played by technological advances and thereby unavoidably forecasting global depletion is to ignore historical data and processes that should be a part of a valid scientific analysis. The world has depended on technological advances in many industries, and to think that they will stop is nonsensical. Forecasts of global oil depletion should not depend on global endowment estimates that have not held up. Predictions should not rely so heavily on the convenient and simplistic projections of historical discovery and production data.

Increased transportation fuel efficiency is essential to reducing future oil demand. In the 1970s, annual global oil production was about 5.1 barrels per person. When the CAFE standards of 27.5 mpg improved new fuel economy by 14 mpg, annual global per capita production dropped to 4.1 barrels per person, and that figure has remained fixed for the past 25 years. Global oil use has grown at the same rate as the increase in population. If fuel efficiency were increased to 43 mpg, oil production might fall to 3 barrels per person per year. At that per capita rate, peak global oil production would occur during the anticipated period of maximum global population of 9.22 billion in 2075.[77] Under that scenario, annual global oil production requirements would be 27.7 billion barrels, which is essentially the global production value seen in the past few years. Should the world of the future choose to use oil as we do today to power transportation, our current rate of production would be sufficient as long as vehicle fuel economy were improved by about 60 percent. This prospect seems likely. Fuel economy improved by more than 60 percent in response to the original 1975 CAFE standards, and many small cars and hybrids already get 43 to 50 mpg (5.5 to 4.7 liters per 100 km). The 2007 revised CAFE standard of 35 mpg by 2020 added 7.5 mpg, but at least another 7.5 mpg gain is needed. The rest of the world is ahead of the US on vehicle fuel economy, and the US must catch up and begin to take the lead. The Obama administration is aiming to shorten the time-line for fuel-economy compliance to 2016 and make the standard 35.5 mpg, roughly in line with the California requirement.[78] This is a move in the right direction, but it is not enough to facilitate a concerted effort to replace light-fleet vehicles with plug-in hybrids and electric cars.

After a viable technology that improves efficiency is introduced, its adoption spreads. Typically, there is no widespread reversion to an older inefficient approach. The global transfer of new technology affects oil recovery and consumption. A rapid increase in oil prices can have a positive effect on technology and the direction of our oil-consumption path. High prices promote efficiency and the introduction of substitutes. As a consequence of innovation, the ultimate result of an increase in price is a sustainable lower price. However, that progression will likely involve coping with oil-supply disruptions and oil-price volatility. Alternatively, a new model would be to actively promote oil-price stability and allow for the orderly transition to the most sensible transportation-fuel alternatives.

If there is a peak and decline in global oil production during the next two decades, it is more likely that it will reflect a decrease in global oil demand, rather than production choked by critically low global availability. The state of global oil resources (listed above) suggests that improvements in technology and efficiency will allow for continued use of conventional oil resources. The line between conventional and unconventional sources of oil will blur as more unconventional sources come on line. However, issues other than availability have become increasingly important to our future use of oil. Driven by security, stability, and environmental concerns, major consuming nations may shift away from conventional oil as a transportation fuel. Based on the history of production of other non-renewable Earth resources, a move away from today's conventional oil will take place long before the end of the global endowment is in sight.

Notes and References

1. US Department of Energy (2008). *Transportation Energy Data Book*, Oak Ridge National Laboratory, Edition 27.

2. US Department of Transportation (2008). Federal Highway Administration, "Vehicle Miles Traveled (VMT) vs. Mobile Source Air Toxics Emissions, 2000–2020," www.fhwa.dot.gov/environment/vmtems.htm and www.afdc.energy.gov/afdc/data/

3. Lovins, A. B. (2006). "Reinventing the Wheels: The Automotive Efficiency Revolution," *Economic Perspectives*, **11**(2), July 2006, http://usinfo.state.gov/journals/ites/0706/ijee/lovins.htm

4. US Environmental Protection Agency (2006). "Light-Duty Automotive Technology and Fuel Economy Trends: 1975 through 2006," EPA420-S-06-003.

5. Buckingham, D. A. (2006). "Steel stocks in use in automobiles in the United States," USGS Fact Sheet 3144.

6. Andrews, E. L. (2007). "Senate Adopts an Energy Bill Raising Mileage for Cars," *The New York Times*, June 22, 2007. The Union of Concerned Scientists estimates that, if implemented, this new fuel economy standard will save just under 6 percent of current US oil consumption, or the equivalent of removing 30 million cars from the road.

7. Fairley, P. (2008). "The New CAFE Standards – Fuel standards will likely be achievable but won't encourage innovation," ABC News, January 15, 2008, http://abcnews.go.com/Technology/GlobalWarming/story?id=4136951&page=1

8. An, F. and A. Sauer (2004). "Comparison of Passenger Vehicle Fuel Economy and Greenhouse Gas Emission Standards Around the World", Pew Center on Global Climate Change.

9. Data courtesy of Jim Nader, Director, North American Business Development, CSM Worldwide, August, 2007.

10. Foster, J. B. (2000). "Capitalism's Environmental Crisis – Is Technology the Answer?" *Monthly Review*, **52**(7).

11. Sorrell, S. (2009). "Jevons' Paradox revisited: The evidence for backfire from improved energy efficiency," *Energy Policy*, **37**(4), April 2009: 1456–69.

12. The 27.5 mpg standard went into effect in 1985 but was reduced from 1986 to 1989 and reinstituted in 1990.

13. Bezdek, R. H. and R. M. Wendling (2005). "Fuel Efficiency and the Economy," *American Scientist*, **93**(2): 132.

14. Greene, D. L., J. Kahn, and R. Gibson (1999). "Fuel Economy Rebound Effect for US Household Vehicles," *The Energy Journal*, **20**(3): 1–31.

15. National Research Council (2002). *Effectiveness and Impact of the Corporate and Fuel Economy Standards*. Washington DC: National Academy Press.

16. Greening, L. A., D. L. Greene, and C. Difiglio (2000). "Energy efficiency and consumption – the rebound effect – a survey," *Energy Policy*, **28**: 389–401.

17. Heavenrich, R. M. (2006). "Light-Duty Automotive Technology and Fuel Economy Trends: 1975 Through 2006," US Environmental Protection Agency, July 2006, EPA420-R-06-011.

18. EPA (2007). "Light-Duty Automotive Technology and Fuel Economy Trends: 1975 through 2007," US Environmental Protection Agency, September 2007.

19. National Highway Traffic Safety Administration (2008). www.nhtsa.gov/cars/rules/CAFE/ImportedCarFleet.htm

20. National Research Council (2002). *Effectiveness and Impact of the Corporate and Fuel Economy Standards*. Washington DC: National Academy Press.

21. United Nations Environment Programme (2008). "Asia and Pacific Vehicle Standards and Fleets," www.unep.org/pcfv/PDF/AsiaPacificVehicleMatrix-June2008.pdf; "In India, Cheap Car Will Challenge Two-Wheelers," National Public Radio (NPR), July 8, 2008, www.npr.org/templates/story/story.php?storyId=92289904; http://data.un.org/CountryProfile.aspx?crname=India

22. US Department of Energy (2007). "Changes in Vehicles per Capita around the World," Fact #474, June 18, 2007; International Monetary Fund (2005). www.imf.org/external/pubs/ft/weo/2005/01/chp4data/fig4_7.csv

23. Kobos, P. H., J. D. Erickson, and T. E. Drennen (2003). "Scenario analysis of Chinese passenger vehicle growth," *Contemporary Economic Policy*, **21**(2): 200–17.

24. *CIA World Factbook* (2009); based on nominal GDP estimates for 2008.

25. Based on a 100,000 rupee sale price in mid-2009. www.tatamotors.com/our_world/profile.php

26. National Research Council, US National Academy of Engineering (2008). *Energy Futures and Urban Air Pollution: Challenges for China and the United States*. ISBN 0-309-11141-2, 386 pp.

27. New York City Department of Transportation (2007). www.nyc.gov/html/dot/downloads/pdf/safetyrpt07_1.pdf

28. McShane, C. (1994). *Down the Asphalt Path: The automobile and the American city*. New York: Columbia University Press, 288 pp.

29. Greene, D. L. and P. N. Leiby (2006). "The Oil Security Metrics Model: A Tool for Evaluating the Prospective Oil Security Benefits of DOE's Energy Efficiency and Renewable Energy R&D Programs," Oak Ridge National Laboratory, ORNL/TM-2006/505; Greene, D. L. (2008). "Oil Security Metrics Model," Oak Ridge National Laboratory, April 2008, www1.eere.energy.gov/vehiclesandfuels/facts/2008_fotw522.html

30. Congressional Budget Office (2009). The Budget and Economic Outlook: Fiscal Years 2009 to 2019.

31. Greene, D. L. and P. N. Leiby (2006). "The Oil Security Metrics Model: A Tool for Evaluating the Prospective Oil Security Benefits of DOE's Energy Efficiency and Renewable Energy R&D Programs," Oak Ridge National Laboratory, ORNL/TM-2006/505; Greene, D. L. (2008). "Oil Security Metrics Model," Oak Ridge National Laboratory, April 2008, www1.eere.energy.gov/vehiclesandfuels/facts/2008_fotw522.html

32. US Department of Energy (2008). *Transportation Energy Book*, Oak Ridge National Laboratory, Edition 27.

33. There are 5.5 pounds of carbon in a gallon of gasoline; a gallon of gasoline weighs 6.3 pounds, of which 87 percent is carbon. On combustion, each pound of carbon bonds with 2.7 pounds of oxygen. Adding the carbon in a gallon of gasoline to the oxygen consumed on combustion, over 20 pounds of carbon dioxide is formed.

34. Government Accountability Office (2007). "Crude oils: Uncertainty about future oil supply makes it important to develop a strategy for addressing a peak and decline in oil production," United States Government Accountability Office, GAO-07-283.

35. Searchinger, T., et al. (2008). "Use of U.S. Croplands for Biofuels Increases Greenhouse Gases Through Emissions from Land-Use Change," *Science*, **319**: 1238.

36. Hill, J., S. Polasky, E. Nelson, D. Tilman, H. Huo, L. Ludwig, J. Neumann, H. Zheng, and D. Bonta (2009). "Climate change and health costs of air emissions from biofuels and gasoline," *Proceedings of the National Academy of Science*, **106**(6), February 9, 2009: 2077–82.

37. Westcott, P.C. (2007). "U.S. Ethanol Expansion Driving Changes Throughout the Agricultural Sector," *Amber Waves*, September, Economic Research Service, US Department of Agriculture.

38. Westcott, P. C. (2007). "Ethanol Expansion in the United States: How Will the Agricultural Sector Adjust?" USDA Report, FDS-07D-01, May 2007.

39. Gustafson, C. (2007). "Biofuels Economics: How Many Acres Will Be Needed For Biofuels?" Part I, News, North Dakota State University Agricultural Communication.

40. Data from www.ers.usda.gov/Data/Feedgrains/FeedGrainsQueriable.aspx and EIA. In 2008, 12,200 million bushels of corn were produced in the US, of which 4,000 million were for corn ethanol. In 2007, 13,100 million bushels of corn were produced in the US, of which 3,000 million were for corn ethanol.

41. Krauss, C. (2009). "Ethanol, Just Recently a Savior, Is Struggling," *The New York Times*, February 12, 2009.

42. Kennedy, D. (2007). "The biofuels conundrum," *Science*, **315**, April 27, 2007.

43. Rae, J. B. (1955). "The Electric Vehicle Company: A Monopoly That Missed," *The Business History Review*, **29**(4), December 1955: 298–311; Goodsell, C. M. and H. E. Wallace (1901). *The Manual of Statistics: Stock Exchange Handbook*, original from the University of Michigan, digitized July 5, 2007.

44. Assuming 25 miles per gallon, a range of 8 to 13 cents per kWh, $2.77 per gallon, and 3 cents per mile for an electric vehicle. Alternative Fuels and Advanced Vehicles Data Center, www.afdc.energy.gov/afdc/vehicles/plugin_hybrids_benefits.html. Comparison based on Uhrig, R. E. (2005). "Using plug-in hybrid vehicles to drastically reduce petroleum-based fuel consumption and emissions," *The Bent of Tau Beta Pi*, Spring: 13–19. A light vehicle getting 20 mpg is 20 percent efficient and would use 0.367 kWh per mile. A plug-in at

70 percent efficiency would use 0.603 kWh per mile. At $3 per gallon, a standard vehicle would cost 15 cents per mile, while a plug-in using 13 cents per kWh would cost 8 cents per mile.

45. Wohlsen, M. (2008). "Tesla Motors to build electric sedan in California," The Associated Press, Wednesday, September 17, 2008, (*Washington Post*); www. teslamotors.com

46. LaMonica, M. (2009). "Tesla Motors CEO: Model S is cheaper than it looks," CNET News, March 27, 2009.

47. Reuters (2009). "Daimler to expand car battery alliance," March 14, 2009.

48. Campbell, J. E., D. B. Lobell, and C. B. Field (2009). "Greater Transportation Energy and GHG Offsets From Bioelectricity Than Ethanol," *Science*, Science-express Report, May 7, 2009.

49. Kang, B. and G. Ceder (2009). "Battery materials for ultrafast charging and discharging," *Nature*, **458**, March 12, 2009, doi:10.1038/nature07853; Thomson, E. A. (2009). "Re-engineered battery material could lead to rapid recharging of many devices," *MIT News*, March 11, 2009.

50. Hadley, S. W. and A. Tsvetkova (2008). "Potential impacts of plug-in electric hybrid vehicles on regional power generation," Oak Ridge National Laboratory, ORNL/TM-2007/150.

51. Uhrig, R. E. (2005). "Using plug-in hybrid vehicles to drastically reduce petro-leum-based fuel consumption and emissions," *The Bent of Tau Beta Pi*, Spring: 13–19.

52. Kuhn, A. (2009). "Chinese Electric Car Jolts The Competition," National Public Radio (NPR), January 13, 2009, www.npr.org/templates/story/story.php? storyId=99286134; Moran, T. (2009). "Dealing With Realities of an Electric-Car Fleet," *The New York Times*, January 22, 2009; Ewing, J. (2009). "Test-Driving the Electric Mini," Business Week, March 12, 2009.

53. US Department of Energy (2009). "Carbon Reduction of Plug-in Hybrid Electric Vehicles," Energy Efficiency and Renewable Energy Fact #562, March 15, 2009, www1.eere.energy.gov/vehiclesandfuels/facts/2009_fotw562.html

54. Elgowainy, A., A. Burnham, M. Wang, J. Molburg, and A. Rousseau (2009). "Well-to-Wheels Energy Use and Greenhouse Gas Emissions, Analysis of Plug-in Hybrid Electric Vehicles," ANL/ESD/09-2, Center for Transportation Research, Argonne National Laboratory.

55. Of all drivers surveyed, 78 percent travel 40 miles or less round-trip from home to work, as reported in Werber, M., M. Fischer, and P. V. Schwartz (2009). "Batteries: Lower cost than gasoline?" *Energy Policy*, **37**: 2465–8, citing US Department of Transportation, Bureau of Transportation Statistics, vol. 3, issue 4, October, 2003.

56. Hadley, S. W. and A. Tsvetkova (2008). "Potential impacts of plug-in electric hybrid vehicles on regional power generation," Oak Ridge National Laboratory, ORNL/TM-2007/150.

57. Krugman, P. (2009). "Empire of Carbon," *The New York Times*, May 14, 2009.

58. Energy Information Administration (2008). *International Energy Outlook 2008*, Chapter 7, Energy-Related Carbon Dioxide Emissions, www.eia.doe.gov/oiaf/ieo/emissions.html

59. Stone, D. (2008). "Blowing Smoke, Is clean coal technology fact or fiction?" Newsweek Web Exclusive, December 9, 2008, www.newsweek.com/id/173086

60. Zakaria, F. (2008). "Al Gore during an October speech on climate change," *Newsweek*, December 8, 2008.

61. US Department of Energy (2009). "Clean Coal Technology Roadmap," National Energy Technology Laboratory, CURC/EPRI/DOE Consensus Roadmap, www.netl.doe.gov/technologies/coalpower/cctc/ccpi/pubs/CCT-Roadmap.pdf

62. Harrabin, R. (2007). "China building more power plants," BBC News, 19 June 2007, http://news.bbc.co.uk/2/hi/asia-pacific/6769743.stm

63. Energy Information Administration (2008). *World Energy Outlook 2008*, Chapter 4, Coal. www.eia.doe.gov/oiaf/ieo/coal.html

64. Shuster, E. (2009). "Tracking New Coal-Fired Power Plants," US DOE National Energy Technology Laboratory, January 5, 2009, www.netl.doe.gov/coal/refshelf/ncp.pdf

65. Rosenthal, E. (2008). "Europe Turns Back to Coal, Raising Climate Fears," *The New York Times*, April 23, 2008.

66. PricewaterhouseCoopers (2009). "How managing political instability can improve business performance," www.pwc.com/extweb/service.nsf/docid/fe926b4a5ce358a5852571d8005f0b12#

67. Transparency International (2009). www.transparency.org/about_us

68. US Department of Energy (2009). "Releasing crude oil from the Strategic Petroleum Reserve," www.fe.doe.gov/programs/reserves/spr/spr-drawdown.html. Note that SPR oil was purchased at an average price of $28.42 per barrel ($17 billion in total).

69. Victor, D. G. and S. Eskreis-Winkler (2008). "In the Tank, Making the Most of Strategic Oil Reserves," *Foreign Affairs*, **87**(4): 70–83.

70. An oil stockpile of 3.8 billion barrels is equivalent to 1.9 years of US imports from OPEC, or 4.4 years of US imports of crude oil from the Persian Gulf. The capital cost of SPR storage is about $3.5 per barrel. If applied to a more extensive oil storage system, the capital cost could be thought of as a supply insurance premium.

71. US Treasury (2009). "US Economic Statistics," monthly data, May 4, 2009.

72. US Bureau of Transportation Statistics (2009). "National Transportation Statistics 2009, Number of US Aircraft, Vehicles, Vessels, and Other Conveyances," www.bts.gov/publications/

73. US Department of Energy (2009). "Cars are growing older," Energy Efficiency and Renewable Energy Fact #567, April 20, 2009, www1.eere.energy.gov/vehiclesandfuels/facts/2009_fotw567.html

74. Cuddington, J. T. and D. L. Moss (2001). *American Economic Review*, **91**(4): 1134–48.

75.　www.spe.org/speapp/spe/jpt/1999/09/frontiers_drilling_tech.htm

76.　www.greatachievements.org/?id=3677

77.　United Nations (2004). "World population to 2300," United Nations Department of Economic and Social Affairs, ST/ESA/SER.A/236, 240 pp.

78.　Broder, J. M. (2009). "Obama to Toughen Rules on Emissions and Mileage," *The New York Times*, May 18, 2009.

Index

air quality, 206
Alaska, 65, 128
   heavy oil, 166
   oil sands, 168
Algeria, 23, 41
aluminum, 105–6, 107
alternatives, oil and energy, 115, 210–2, 214–9
Angola, 23, 142
anticlinal traps, 140
Arab–Israeli war (1973), 63, 115
Arctic region, oil and gas, 146
Association for the Study of Peak Oil, 88, 124
Athabasca sands, 168–9
available oil, 119
Azadegan oil field, 138

Bahrain, 23
Baker Hughes drill-rig count, 141–2
Baku, 160
barrels, origin as measure, 20
batteries, 108, 214–15, 228
battleships, 62
bbl, barrel, origin, 20, 52
bell-shaped curve *see* logistic curve
Berman, Arthur, 144
bets, 103–4
biodiesel, 212–13
biofuels, 210–13
bitumen, 168

Bohai Bay oil field, 138–9
Bolivia, oil reserves, 144
booking of oil reserves, 125–6
BP
   coal endowment estimate, 180
   oil price, 77–8, 115–6, 154
   oil production, China and India, 76
   oil production decline data, 65
   oil reserves, 23, 126
Brazil
   oil reserves, 144
   oil shale, 172
Brent Blend, 41
British Petroleum *see* BP
Brookhart, Maurice, 174
Brown, Harrison, 106
Buffett, Warren, 215
Burgan Greater oil field, 71

CAFE standards, 197–9
   effects, 199–205, 223–4
California
   gold, 156, 157
   heavy oil, 166
   oil sands, 168
California Energy Commission, 47
Campbell, Colin, 124
Canada
   oil reserves, 122, 132–3
   oil sands, 27, 29, 122, 132, 136, 168–70